Child Restraint Use Survey

LATCH Use and Misuse

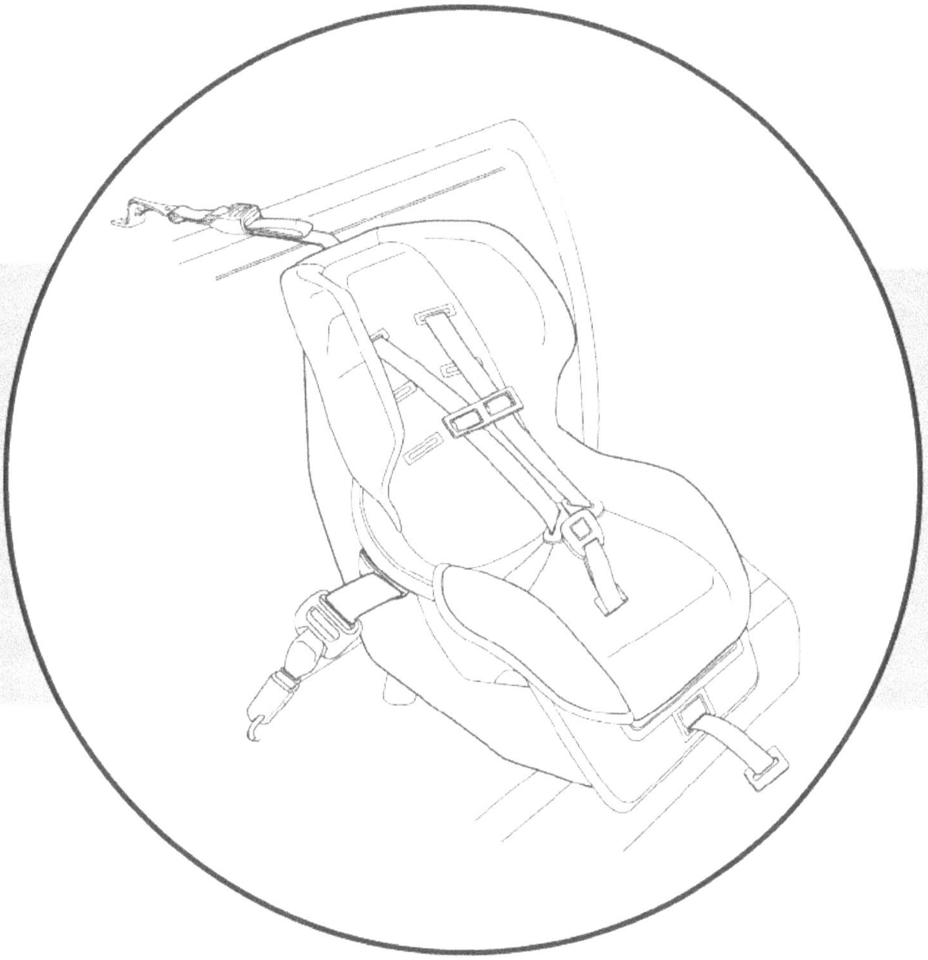

www.nhtsa.gov

NHTSA's National Center for Statistics & Analysis

1. Report No. DOT HS 810 679	2. Government Accession No.	3. Recipient's Catalog No.
4. Title and Subtitle Child Restraint Use Survey: LATCH Use and Misuse		5. Report Date December 2006
		6. Performing Organization Code
7. Author(s) Lawrence E. Decina, Kathy H. Lococo, and Charlene T. Doyle		8. Performing Organization Report No.
9. Performing Organization Name and Address TransAnalytics, LLC 1722 Sumneytown Pike, Box 328 Kulpsville, PA 19443 Evaluation Division; National Center for Statistics and Analysis National Highway Traffic Safety Administration Washington, DC 20590		10. Work Unit No. (TRAIS)
		11. Contract or Grant No. DTNH22-03-C-06010
12. Sponsoring Agency Name and Address National Highway Traffic Safety Administration U.S. Department of Transportation NPO-131, 400 Seventh Street SW. Washington, DC 20590		13. Type of Report and Period Covered NHTSA Final Report
		14. Sponsoring Agency Code
15. Supplementary Notes		

NHTSA conducted a survey from April to October 2005 to collect information about the types of restraint systems that were being used to keep children safe while riding in passenger vehicles. In particular, NHTSA was interested in whether drivers with Lower Anchors and Tethers for CHildren (LATCH)-equipped vehicles were using LATCH to secure their child safety seats to the vehicle, and if so, were these seats properly installed. The make/model and the type of restraint installed in each seating position were recorded for each of the vehicles; demographic characteristics and the type of restraint system were collected for each occupant. In addition, information was gathered about the drivers' knowledge of booster seats and LATCH, along with their opinions on how easy it was to use LATCH.

A key finding of the survey was that 55 percent of child safety seats, located in a seating position equipped with an upper anchor, were attached to the vehicle using an upper tether. Other findings include: (1) In 13 percent of the observations, the child safety seat was placed in a seat position in the vehicle not equipped with lower anchors - the seat belt was used to secure the child safety seat to the vehicle. (2) Among the 87 percent who do place the child safety seat at a position equipped with lower anchors, 60 percent use the lower attachments to secure the child safety seat to the vehicle. (3) 81 percent of upper tether users and 74 percent of lower attachments users said upper tether and/or lower attachments were easy to use. (4) 75 percent preferred lower attachments over seat belts of those with experience using both lower attachments and seat belts. (5) 61 percent of upper tether nonusers and 55 percent of lower attachments nonusers cited their lack of knowledge – not knowing what they were, that they were available in the vehicle, the importance of using them, or how to properly use them - as the reason for not using them.

17. Key Words Lower Anchors and Tethers for Children, LATCH, Child Restraint Systems, Use and Misuse, Data Collection Procedures, Research Methodology		18. Distribution Statement No restrictions. This document is available to the public through the National Technical Information Service, Springfield, VA 22161.	
19. Security Classif. (of this report) Unclassified	20. Security Classif. (of this page) Unclassified	21. No. of Pages 113	22. Price

Form DOT F 1700.7 (8-72)   Reproduction of completed page authorized

TABLE OF CONTENTS

TABLE OF CONTENTS
(Continued)

LIST OF TABLES

LIST OF TABLES
(Continued)

EXECUTIVE SUMMARY

Lower Anchors and Tethers for Children (LATCH), an installation system created to help standardize the way child restraints are attached to vehicles without using a seat belt, has been in the marketplace since the NHTSA-promulgated regulation (Federal Motor Vehicle Safety Standard 225) became fully effective on September 1, 2002. All child restraints and most new vehicles manufactured as of that date were required to include hardware components designed to simplify child safety seat installation and to reduce the continuing high incidence of misuse and incorrect installation of child safety seats.

As is common with most new technologies, issues have arisen from both manufacturing and real life experiences. This study provides a detailed look into the use of LATCH, information on consumers' knowledge and attitudes about it, and steps aimed toward broadening the correct use of this new system.

Upper tethers for child safety seats reduce the tilting or rotation of the seat during a frontal crash, because they provide a second attachment of the seat to the vehicle, when used with a seat belt or lower anchors. Sled testing using forward-facing child safety seats clearly demonstrates that tethered seats, when correctly used, tend to reduce occupant injury levels when compared to the same seat without the upper tether attached. Whereas some safety seats were equipped with upper tethers as early as 1970, use of upper tethers remained negligible, in part because many vehicles did not have upper tether anchors. In response, NHTSA established Federal Motor Vehicle Safety Standard (FMVSS) No. 225 – Child Restraint Anchorage Systems – to require upper tether anchorages in the back seats of passenger vehicles effective September 1, 2000. The agency amended FMVSS No. 213 – Child Restraint Systems – to require an upper tether on child safety seats effective September 1, 1999.

The traditional method of attaching child safety seats with seat belts was prone to misuse such as a loose fit or incorrect routing. The new LATCH system, which has a user-friendly system of two lower attachments on the safety seats that connect with lower anchors built into the back seats of passenger vehicles, was developed as an alternative method of attaching child safety seats to the vehicle. FMVSS No. 213 required the lower attachments on child safety seats, effective September 1, 2002, and FMVSS No. 225 phased lower anchors into new vehicles by the same date.

Between April and October 2005, NHTSA conducted an observational survey of the use, misuse and consumer reaction to LATCH. Data was collected at 66 sites – shopping centers, child care facilities, health care centers, and recreation facilities – in 31 counties in Arizona, Florida, Michigan, Missouri, North Carolina, Pennsylvania, and Washington. For 1,121 children from birth to age 4 in child safety seats riding in the back seat of the vehicle, we observed the vehicle and occupant restraint equipment available for each seating position in the vehicle and how the safety seats were installed in the vehicle. In addition, we interviewed the drivers on their satisfaction with LATCH and their reasons for using or not using it. This survey focused on restrained child passengers of the relatively new, LATCH-equipped vehicles; it was not designed to monitor the overall use rate of child safety seats in the United States.

A key finding of the survey was that 55 percent of child safety seats, located in a seat position equipped with an upper anchor, were attached to the vehicle using an upper tether when

1

appropriate. This is a substantial improvement over the 15 percent use of upper tethers on the child safety seats equipped with upper tethers during the 1980s. But many parents are not yet protecting their children with this technology, even though the use of an upper tether reduces the forward motion of the safety seat in a crash; thereby, reducing the potential for injury-causing contacts with vehicle interior surfaces, as well as stress and injury to the vulnerable head/neck/spinal cord area of the body.

Other principal findings and conclusions from the survey include:

USE RATE FOR LOWER ATTACHMENTS AND ANCHORS

- In 13 percent of the observations, the child safety seat was placed in a seat position in the vehicle not equipped with lower anchors; the seat belt was used to provide the lower attachment for the child safety seat.

    o Almost all these children were in a center-rear seat, the position long recommended to parents because of its distance from potential points of impact. Although FMVSS 225 requires lower anchors at two rear-seat positions, the center-rear seat is equipped with lower anchors in only a relatively few vehicles because, in most vehicles, the anchors are located in the outboard (window) seats.

- Among the 87 percent who do place the child safety seat at a position equipped with lower anchors:

    o Sixty percent use the lower attachments and anchors to provide the lower attachment for the seat, including:

        ▪ Forty-seven percent using only the lower attachments and lower anchors; and

        ▪ Thirteen percent using the lower attachments and lower anchors plus the seat belt.

    o Forty percent of parents continue to rely on the vehicle's seat belts for installing the safety seat.

SECURE INSTALLATION OF CHILD SAFETY SEATS

- Sixty-one percent of child safety seats installed with lower attachments and anchors were securely installed.

- In a 2002 survey of child safety seats installed with seat belts, only 40 to 46 percent of the child safety seats were securely installed.

- Only 11 to 28 percent of child safety seats were securely installed with seat belts in a 1995 survey.

- Conclusion: lower attachments and anchors furnished by LATCH appear to be helping to reduce the insecure installation of child safety seats.

CONSUMER REACTION TO LATCH

- Eighty-one percent of parents using the upper tether and anchor said they were easy to use.

- Seventy-four percent using the lower attachments and anchors said they were easy to use.

- Seventy-five percent of parents with experience using both lower attachments and anchors, as well as seat belts to secure child safety seats, said they preferred lower attachments and anchors over seat belts.

- Conclusion: people who have experience with LATCH appear to prefer its use over the conventional attachment method using the vehicle seat belt.

REASONS FOR NOT USING LATCH

- Sixty-one percent of parents who did not use the upper tether cited their lack of knowledge as the reason for nonuse; they did not know what a tether was, they did not know that a tether system was available in their vehicle, they did not know the importance of using a tether, or they did not know how to properly use a tether.

- Fifty-five percent of parents who did not use the lower attachments cited their lack of knowledge as the reason for nonuse.

- Conclusion: The primary reason people do not use LATCH is that they don't know about it, although the absence of lower attachments and anchors in some center-rear seats is another reason cited for not using the LATCH system.

LONG-TERM GOALS

Based on a review of the state-of-the-art in child passenger safety as well as the survey results, NHTSA believes the following goals would be beneficial to safety:

- Increase proper child safety seat installation, whether by using the LATCH system or, in some cases, by using seat belts in the center-rear or other seating positions.

- Increase the use of the upper tether strap and anchor, because tether use has safety benefits. This means that the upper tether strap and anchor should be used whether using the lower anchors or the seat belt system to secure the child restraint in a center-rear seat.

- Increase the percentage of children riding in the center-rear seat, which is the safest place in a vehicle, in those cases in which the child safety seat can be securely installed in the center-rear seat. This includes whether the child safety seat is attached to the vehicle with lower anchors or a seat belt.

- Increase the use of lower attachments and anchors, when they are available, because lower attachment use helps enhance the correct installation of safety seats.

- Increase the proportion of new vehicles that have lower anchors in center-rear seats, since the lower anchors are used correctly more often than installations using seat belts.

- Explore ways to increase the number of children who are safely restrained, for the population of children who weigh over 40 pounds but are not yet old enough (at least 4 years old) to ride in booster seats. This includes ways to increase the use of upper tether straps and upper tether anchors, in combination with child safety seats that are weight/age appropriate.

NEXT STEPS

- Convene a meeting with stakeholders, including automakers, safety seat manufacturers, leading researchers and prominent national non-profit organizations to discuss issues related to educating and supporting consumers in their efforts to use LATCH to safely restrain their children and any other efforts the Agency could take to improve proper installation and use of child safety seats.

INTRODUCTION

This section provides the background for the study, including the definition of LATCH (Lower Anchors and Tethers for Children), as well as the project objectives. A glossary of child restraint terms is provided in Attachment A of this report.

BACKGROUND

Child safety seats, also called child restraint systems (CRSs) are the most effective way to protect young children involved in motor vehicle crashes from serious injury or death. The National Highway Traffic Safety Administration estimates that CRSs, when properly used, reduce the chance of death in a motor vehicle crash by 71 percent.[1]

CRSs vary according to the size of the child they are designed to restrain, the direction the child should face, the type of internal restraining system, and the method of installation. CRSs are designed to provide two links between the vehicle and the child. The CRS is securely attached to the vehicle seat using the vehicle seat belt or the lower attachments, at the same time that the child is properly secured in the CRS with a separate harness and/or other restraining surface. These two links between the vehicle and the child are critical in reducing injuries or death in the event of a vehicle crash.[2]

While a 2002 NHTSA CRS misuse observation study showed improvement in the overall restraint use for young children, there were still specific types of CRS misuse that were common, especially loose harness straps restraining the child in the CRS and a loose connection between the CRS and vehicle seat.[3] Child passenger safety experts believe that a loose CRS in a vehicle is a serious misuse and that the LATCH system can simplify CRS installation and reduce misuses of this type.[4]

NHTSA has recognized the difficulties that parents and caregivers experience in securely attaching a CRS to a vehicle and has addressed this issue in 1999 by establishing a uniform child restraint attachment system known as LATCH by amending FMVSS 213, "Child Restraint Systems" and issuing FMVSS 225, "Child Restraint Anchorage Systems," effective September 1, 2002.[5]

This field observation study provided a look into LATCH system use and misuse among the motoring public. As part of the LATCH system, the upper tether provides an additional point of attachment to the vehicle, making it a very important improvement in occupant protection. While the use of lower anchors can be a convenience and possibly improve tightness of fit, the use of an upper tether has been shown to actually reduce the forward movement of the

---

[1] Starnes, M., and Eigen, A., *Fatalities and injuries to 0-8 year old passenger vehicle occupants based on impact attributes,* NHTSA Technical Report No. DOT HS 809 410, National Highway Traffic Safety Administration, Washington, March 2002.

[2] Weber, K., "Crash Protection for Child Passengers – A Review of Best Practice," *UMTRI Research Review,* Vol. 31, No. 3, July-September 2000, pp. 1-27.

[3] Decina, L.E., and Lococo, K., *Misuse of Child Restraints.* NHTSA Publication No. DOT HS 809 671, National Highway Traffic Safety Administration, Washington, 2004.

[4] Stewart D., Lang, N.J., and Emery, S., *LATCH –Lower Anchors and Tethers for Child Restraints*, Safe Ride News Publication, Seattle, 4th Edition, 2005.

[5] *Code of Federal Regulations*, Title 49, Government Printing Office, Washington, 2005, Parts 571.213 and 571.225.

occupant's head, minimizing the risks of impact of the child's head with the vehicle interior or other passengers. Thus, LATCH holds the promise of not only simplifying the installation of child restraint systems and minimizing misuse, but has the potential to improve child occupant protection.

LATCH-equipped vehicles have at least two sets of small bars, called anchors, located in the back seat where the seat cushions meet. LATCH-equipped CRSs have a lower set of attachments that fasten to these vehicle lower anchors. Most forward-facing CRSs also have a top strap (upper tether) that attaches to a top or upper anchor in the vehicle. Together, they make up the LATCH system.[6] It is believed that measurable improvement in the proper and secure use of CRSs can be achieved, based upon the premise that this new LATCH system may be more convenient and easier to use than vehicle seat belts.[7]

While LATCH holds the promise of simplifying the installation of CRSs, drivers transporting young children still must manually connect the CRS's tether strap to the vehicle anchor behind the seat and connect the CRS lower attachments to the vehicle's lower anchor bars. Once connected, the upper tether and lower attachment straps need to be tightened. In addition, there are exceptions to LATCH connections based on the make/model of the type of the CRS and the vehicle make/model's seat belts and lower anchors. With all of these circumstances, it is not surprising that some parents and caregivers transporting young children have had difficulties initially installing their CRSs using LATCH, as evidenced by field observations. The 2002 CRS misuse observation study found LATCH misuse associated with upper tether and lower anchor connections.[8] More recently, a workshop with the same field observation staff provided anecdotal reports of LATCH misuse. They reported misuse associated with tethers wrapped around head restraints; tethers swinging loose in the vehicles; improper connections with embedded lower anchor bars; and lower attachments connected to cargo hooks, seat material, and other structures in the vehicle.[9]

PROJECT OBJECTIVES

The main objectives of this project were to collect quantitative data and other information concerning the restraint use of children in passenger vehicles and to identify the use and misuse of LATCH. The research sought to obtain additional information about drivers' socio-demographic characteristics and their knowledge of restraint systems (booster seat, LATCH issues and laws), as well as characteristics about their vehicles (e.g., seating configurations, restraint and LATCH systems). The task activities required in the scope of work, in order to carry out these objectives are contained in Attachment B.

---

[6] Starnes, M., and Eigen, A., *Fatalities and injuries to 0-8 year old passenger vehicle occupants based on impact attributes*, NHTSA Technical Report No. DOT HS 809 410, National Highway Traffic Safety Administration, Washington, March 2002.

[7] *Final Economic Assessment, FMVSS No. 213, FMVSS No. 225, Child Restraint Systems, Child Restraint Anchorage Systems*, Washington, 1999.

[8] Decina, L.E., and Lococo, K., *Misuse of Child Restraints*. NHTSA Publication No. DOT HS 809 671, Washington, 2004.; Decina, L.E., and Lococo, K., "Child Restraint System Use and Misuse in Six States," *Accident Analysis and Prevention, Vol.* 37, 2005, pp. 583-590.

[9] Decina, L.E., Lococo, K., and Block, A., *Misuse of Child Restraints: Results of a Workshop to Review Field Data Results*, NHTSA Traffic Safety Facts – Research Note No. DOT HS 809 851, National Highway Traffic Safety Administration, March 2005.

RESEARCH METHODOLOGY

This section of the report identifies the research methodology used in meeting the project objectives. Key research task activities centered around developing a survey plan, collecting data, and developing a database, as well as the preparation of a coding manual and production of a SAS-readable data file.

To note, restraint use for children was identified in this study as either a child restraint system (CRS), a booster seat, a vehicle seat belt (either a lap and shoulder belt, a lap belt only, or a shoulder belt only), or unrestrained. The basic types of occupant restraints for children are: infant-only seats, convertible seats (converts from rear facing to forward facing), forward-facing only seats, combination seats (converts from forward facing to booster seat when the harness is removed), booster seats, and integrated (built-in) seats. (There are also less common CRSs in use including car beds for low-birth-weight infants and infants with medical conditions; harness vests for toddlers and older children; and restraint systems for children with special needs.) Some child passenger safety (CPS) experts do not consider booster seats as child restraint systems. This is because booster seats are considered to be devices that raise the child's seating height so that the vehicle lap and shoulder belts fit better. The vehicle belts actually restrain the child, not the booster seat. In this study, the term CRS is reserved for child restraints with an internal harness; it <u>does not</u> include booster seats. A forward-facing combination seat that is being used with the internal harness was considered a CRS; if the internal harness was removed and the seat was being used as a belt-positioning booster, it was considered a booster.

In this study, CRS installations using either the upper tether anchor and/or the lower anchors were observed, and misuses with the LATCH system were recorded. Additionally, each child's weight was obtained, so that upper tether and lower anchor use by weight could be analyzed.

For upper tether anchor use and misuse the following elements were studied:
- Presence of tether anchor in vehicle;
- Presence of upper tether straps on CRS;
- CRS upper tether adjustor type (tilt-lock, double-back);
- Whether the upper tether was being used;
- Whether a combination seat used as a belt-positioning booster was using the upper tether;
- Tether routing;
- Whether an upper tether was attached to a designated upper tether anchor or to some other place; and
- Tether twists; and snugness of tether attachment.

For lower anchor use and misuse, the following elements were studied:
- Presence of lower anchor in vehicle seat position;
- Presence of lower attachments on CRS;

- Lower attachment connector type (flexible strap, hook-on or push-on or rigid), flexible strap release type (squeeze or tilt-lock), and strap type (single, side straps, two straps);
- Whether the lower anchors were in use;
- Whether the vehicle's seat belt was used to fasten the CRS in addition to or instead of the lower attachments;
- Whether a combination seat being used as a belt-positioning booster was using the lower attachments;
- Whether the lower attachments were connected to the designated anchors;
- Whether one connector was attached per anchor bar;
- Whether the connector was installed right-side up;
- Whether the lower attachment twists;
- The lower attachment snugness; and
- Whether the correct angle was achieved for infant seats.

SURVEY PLAN

The survey plan incorporated data collection techniques necessary to efficiently and effectively capture the target sample (primarily, children less than 5 years in a CRS in the back seat of a vehicle equipped with upper tether anchors).

Sample Estimates

Based on the 1999 NHTSA rulemaking (FMVSS 213, Child Restraint Systems and FMVSS 225 Child Restraint Anchorage Systems), NHTSA provided an estimate of the number of vehicles on the road with upper tether anchors and lower anchors. Projections for new vehicles and CRSs equipped with LATCH technology were based on the following phasing-in schedules:

For upper tether anchors on vehicles:

- 80 percent of new cars manufactured between September 1, 1999, and August 31, 2000; and
- Nearly 100 percent of new cars and light trucks manufactured on or after September 1, 2000.

For lower anchor points:

- 20 percent of all vehicles manufactured between September 1, 2000, and August 31, 2001;
- 50 percent of all vehicles manufactured between September 1, 2001, and August 31, 2002; and
- Nearly 100 percent of all vehicles manufactured on or after September 1, 2002.

NHTSA provided preliminary recommendations for survey size for each group of child passengers identified for the study. However, it was determined that data collection could end once the target of 1,000 child passengers in Group 3 had been reached. The Groups were:

Group 1: Child passengers age 0 to 12 (approximately 6,750);

Group 2: Child passengers age 0 to 4 years old in a CRS in the back seat (approximately 2,250);

Group 3: Child passengers age 0 to 4 years old in a CRS in the back seat of a vehicle equipped with upper tether anchors (approximately 1,000);

Group 4: Child passengers age 0 to 4 years old in a CRS in the back seat of a vehicle equipped with lower anchors and upper tether anchors (approximately 585).

The groups are not mutually exclusive; rather, they are subsets. A Group 4 child is also part of Groups 3, 2, and 1. A Group 3 child is also part of Groups 2 and 1. A Group 2 child is also part of Group 1. Also, it is important to note that the definition of a Group 3 or Group 4 child did <u>not</u> require that the child was riding in a <u>seating position</u> equipped with lower anchors and/or an upper tether anchor; it only required that the child was <u>riding in a vehicle equipped with lower anchors and/or upper tether anchors</u>.

Office of Management and Budget (OMB) Review

Data collection forms were developed for the Child Restraint Use Survey. The forms were based upon meeting the project goals; past NHTSA child observation survey forms and experience; and input by the State site coordinators (SSCs) and other national CPS experts. Supporting documentation was prepared for use in requesting the Office of Management and Budget's (OMB) review and approval of these data collection forms. See Attachment C for details on the material prepared and submitted to OMB.

Site Selection Criteria

A purposive or convenience sample technique was used for the study. Data collection sites met the following criteria:

- Socio-demographic and economic area that would have a high probability of newer vehicles with LATCH systems.

- Locations that attract a high frequency of vehicles/drivers with young children (e.g., shopping centers, pediatric centers, amusement parks, fast-food restaurants).

- Locations suitable for safe and efficient data collection (e.g., limited entrance ramps, adequate parking lanes, designated pull-off areas).

- Locations with merchant, community, and law enforcement cooperation.

Initial guidelines provided by NHTSA specified that at least six States should be selected, each representing a different area of the country. The guidelines also recommended that observations be made in at least five counties in each State and four to six sites within each county (depending on the size and population of the county). In addition, they suggested that two of the counties in each State be in standard metropolitan statistical areas (SMSA) and the other three counties in non-SMSA areas. The following seven States were chosen, based on these contract requirements, and on past experience with the project team collecting CRS use and misuse data: Arizona, Florida, Michigan, Missouri, North Carolina, Pennsylvania, and Washington. Scheduling and resources determined the ability of the project teams in each State to meet the geographical distribution criteria set forth in the project's goals. A listing of the counties and cities in each State in which data was collected is contained in Attachment D.

Observer and Interviewer's Responsibilities

The project's first priority was to recruit a qualified team of data collectors. State site coordinators (SSCs) who had been selected because of their many years of expertise in child passenger safety, often including membership in organizations dedicated to child passenger safety, were given this responsibility for the state in which they were working. Observers and interviewers were selected based on their expertise and experience in Child Passenger Safety with many of the field crew having previous experience as data collectors on NHTSA's "Misuse of Child Restraint" Study.[10]

The following characteristics were sought for data collectors:

- Certification as a technician or instructor in CPS (via the National Standardized Child Passenger Safety Training Program);
- Experience with types of child restraints and LATCH systems;
- Experience in observing and collecting CPS survey data;
- Flexible with work schedule;
- Physically able to meet demands of field work; and
- Capabilities in reliability, self-motivation, and resourcefulness.

Observer and interviewer responsibilities were as follows:

Observer Responsibilities:

- After receiving driver permission, open the vehicle doors;
- Observe and record LATCH use and misuse data;
- Observe and record other CRS use;
- Observe and record vehicle seating configuration;
- Observe and record restraint types in the vehicle for each seating position; and
- Observe and record the presence of upper tether and lower anchors for each seating position.

---

[10] Decina, L.E., and Lococo, K., *Misuse of Child Restraints*. NHTSA Publication No. DOT HS 809 671, National Highway Traffic Safety Administration, Washington, 2004.

Interviewer Responsibilities:

- Identify the target vehicle entering the designated area;
- Stop the target driver/vehicle with young children;
- Request permission to conduct the survey;
- Introduce themselves and the observer;
- Ask the driver demographic and interview questions;
- Collect other occupant information (age, restraint use);
- Record data; and
- Thank the driver.

The data collectors were given permission to inform drivers of their concern about the misuse or lack of restraint use in the vehicle. These drivers were given literature describing proper restraint use for young children. In many cases, the drivers were given the opportunity to make an appointment to get their CRSs checked at a fitting station or attend a check-up event with the SSCs' organizations.

Data Collection Procedures

Data was collected by teams of two, which consisted of an interviewer and an observer. Interviewer and observer responsibilities were previously described. The field procedures used to collect the data were as follows:

- Select a target vehicle entering the site and approach the driver. Priority was given to late-model vehicles with young child occupants in CRSs in the back seat.

- Identify oneself, briefly explain the purpose of the study (including informing the driver that the children would not be touched or removed from their child restraint systems or from their seating position), and request permission to conduct the observation.

- Upon receipt of driver's permission, direct driver to move vehicle to designated safety zone or parking spot.

- Interviewer asks questions; and observer conducts observations; both record data, respectively.

- Upon completion of interaction, thank the driver.

- Review data collected for all interview and observation forms.

- Move back into position to wait for the next vehicle.

Each site had a field manager responsible for overseeing the field operation. Duties included: observing techniques used by interviewers and observers; supplying data collection forms; collecting the forms; managing staff scheduling; collecting and checking timesheets; and reporting to the SSCs and principal investigator (PI). In many cases, the field managers were also data collectors. Questions about data were brought to the attention of the data collectors. Data was sent by mail (courier service with tracking labels) to the PI on a regular basis. Data was collected from April to October 2005 in seven States (Arizona, Florida, Michigan, Missouri, North Carolina, Pennsylvania, and Washington).

Pilot Testing

A pilot test was conducted to evaluate the efficiency and effectiveness of the data collection procedures and data collection forms (interview, observation). The pilot study was conducted during the spring of 2005 in the Pennsylvania cities of Hershey (Dauphin County) and Bethlehem (Northampton County). The studies were conducted at three types of sites (shopping centers, street checkpoints with law enforcement assistance, and pediatric centers).

The results of the pilot study yielded minor revisions to the data collection forms related to the arrangement of the coding variables. There were no changes to the data collection procedures. However, the use of law enforcement officers to assist in identifying the appropriate target groups at specific vantage points at sites (such as shopping center parking lanes) was considered strongly advantageous to efficient data collecting.

Another objective of the pilot study was to estimate the level of effort necessary (in terms of time and staff resources) to collect data from the targeted drivers with young child occupants in vehicles with LATCH systems. Despite the fact that it would take less than 10 minutes to complete an observation (and survey) with a target vehicle, the number of target vehicles entering a site was not that common. Results showed that Group 3 data (upper tether anchor in vehicle with child less than 5 years old in the back seat in a CRS) could be collected by a team of two at the rate of one observation per 1.25 hours. This finding confirmed the originally proposed estimate, so that it was determined that no staffing changes needed to be made.

Training

Training consisted of three segments: (1) workshop with SSCs; (2) teleconference with SSCs; and (3) in-State training sessions with SSC as instructor (April - June 2005). The SSCs occasionally had questions, interesting findings, or techniques about what they were uncovering during field observations. This information was shared with all of the SSCs through e-mail distribution from the PI.

The workshop conducted at the beginning of the project (March 2004) provided NHTSA and the PI with the opportunity to provide necessary information to the SSCs regarding the type of data that was needed to meet the project's goals. The session provided discussion about definitions of variables, variables to include on each data collection form, and training expectations.

For the teleconference training session (April 2005), each SSC was given a copy of the Training Manual (See Attachment E of this report) in advance to review. The PI spent several hours with the SSCs during the teleconference, in order to review the contents of the manual and the instructions on how to code the survey and the observation forms. During the teleconference, discussions were held concerning how long the field workers training sessions should be, as well as what material should be covered. A decision was made to have 1.5 days of classroom training to be followed first by 1 day in the field, and then by 1 day of close supervision.

For in-State training (April - June 2005), all data collectors were given a training manual, as well as classroom and in-the-field training. Since the data collectors were already CPS trained, training need only to focus on the specifics of this survey and not the broader area of child passenger safety in general. Classroom sessions covered the following topics: introduction and project objectives; definitions of LATCH use and misuse; definitions of other words/terms on the forms; review of interview questions; review of how to record information on the forms, review of interaction with the public; general data collection procedures, sign out procedures; intervention activities (e.g., setting up a check-up appointment or distributing education material); checking data (including use of the "green" book - LATCH manual); and general miscellaneous field work issues (e.g., supplies, incentives, safety gear).

Classroom sessions were followed with practice trials in parking lots using drivers and young children in mock situations. Vehicles with various upper tether anchor and lower anchor configurations were included in the practice. After a day of field practice, data collectors were taken to field sites (mostly shopping centers) to conduct surveys. They were closely supervised by the training staff for at least another day or until staff were comfortable with their data collection activities. Each SSC conducted its own training session with their staff in a classroom setting, then out in the field.

Data Collection

Approximately 50 CPS-certified observers and interviewers participated in the data collection effort covering 66 sites (i.e., shopping centers, child care facilities, health care centers, recreation facilities) in 31 counties across seven States (Arizona, Florida, Michigan, Missouri, North Carolina, Pennsylvania, and Washington) between April and October 2005. Observations and interviews were conducted with approximately 1,200 drivers transporting young child occupants in the vehicles. Most of these vehicles had either upper tether anchor systems or both upper tether and lower anchor systems. Most of these vehicles had children less than 5 years old in the back seat either in a child restraint system or booster seat. Due to the focus of the survey (i.e., to collect data on LATCH use and misuse), a full range of CRSs and safety seat users was not surveyed. Instead, newer model vehicles in more urban/affluent areas were selected in order to obtain information on the population of interest—vehicles equipped with the LATCH system.

Observation data was collected on child occupants, including type of restraint used and a gross measure of proper use versus misuse, as well as vehicle LATCH installation use and misuse (specifically for children less than 5 years old in a CRS). Children remained in their restraints during the observations. Restraint types available for all vehicle seating positions were also recorded. The two forms that were used to collect observation data were:

- NHTSA 1002 – "CRS and LATCH Use Observation Form."

- NHTSA 1002C – "Arrangement of Vehicle-Seating Positions and Occupant Restraint Equipment Available for Each Seating Position."

In addition, interview data were collected. Drivers were asked a few demographic questions about themselves and their children, and their knowledge of booster seats and LATCH systems. If the LATCH systems were being used, drivers were asked questions about the reasons for use, ease of use, and choice over seat belts. Demographic information and restraint type used were also obtained for all other vehicle occupants. The two forms that were used to collect interview data were:

- NHTSA 1002A (a two-sided form) – "CRS and LATCH Use Interview Form."

- NHTSA 1002B – "CRS and LATCH Use Interview Form, Question 6: Occupant Characteristics Chart."

Copies of the data collection forms are presented in Appendix B of the *Training Manual*, at the end of this report (In Attachment E).

DATABASE DESIGN

This section describes the assignment of children into the survey groups, data entry procedures and checks, and the development of the database and coding manual.

Coding

Survey Group Assignment. There are four survey groups covered in the study:

- Group 1 = Child age 0 to 12.
- Group 2 = Child age 0 to 4, in the back seat in a CRS.
- Group 3 = Child age 0 to 4, in the back seat in a CRS, in a vehicle equipped with an upper tether anchor.
- Group 4 = Child age 0 to 4, in the back seat in a CRS, in a vehicle equipped with lower anchors and upper tether anchors.

The groups are not mutually exclusive; rather, they are subsets. A Group 4 child is also part of Groups 3, 2, and 1. A Group 3 child is also part of Groups 2 and 1. A Group 2 child is also part of Group 1.

Child occupants age 12 and under were coded using the highest-numbered category in which they fit. For example, a Group 4 child, who would also be considered a Group 1, Group 2, and Group 3 child, was coded as Group 4. A child coded as Group 1, would be older than age 4, and/or would be sitting in the front seat or sitting in the back seat, but not in a child restraint system (CRS) in the back seat. A child age 0-4 in the front seat, in or out of a CRS, would also be coded as Group 1.

In this study, CRSs do not include booster seats. A booster seat raises the child so the vehicle lap and shoulder belts fit better. The vehicle seat belt restrains the child, not the booster seat. CRSs are defined as child restraints with internal harnesses (infant, rear-facing convertible, and forward-facing with a harness). Children in belt-positioning booster seats (BPBs) were identified as in booster type 5 and were placed in Group 1, because belt-positioning booster seats (and combination seats with the harness removed and used in the booster mode) should not be attached to the vehicle using LATCH technology. Most vehicle manufacturers list a maximum weight of 40 pounds for children in upper tethered restraints. Vehicle manufacturers currently limit the use of their lower anchors to restraints for children weighing up to 48 pounds. Also seat manufacturers state that BPBs should move freely with the occupant during a collision to prevent submarining, which could happen if a BPB were attached using the lower anchors, and the child was restrained with the seat belt. Children in integrated CRSs (i.e., CRSs supplied by the manufacturer and built into the vehicle's seat) were placed in Group 1.

An exception to this coding rule was made for the subset of eight children who were riding in a vehicle with lower anchors, who were restrained in combination seats used (improperly) as both a forward-facing CRS (CRS type = 4) and a belt-positioning booster (Booster type = 6).[11] These children were restrained with the vehicle seat belt and the internal harness, even though the seat was also anchored to the vehicle using LATCH attachments. These children were coded as riding in boosters (restraint type =2), in belt positioning boosters with LATCH attachments (booster type =6), and as Group 4 children (age 0 to 4 in the back seat in CRSs, in a vehicle equipped with lower anchors and upper tethers), to preserve the fact that they were using the vehicle seat belts as a restraint, without excluding them from analyses of target children in CRSs (as they were also using the internal harness).

Data Entry

Accuracy of data entry was maximized by training of project staff and use of templates in data entry. Following data entry, further data-checking (e.g., range checks, skip pattern checks, data consistency checks, review of text items) was conducted. For additional details on data entry procedures and checks, see Attachment F.

Coding Manual

A *Coding Manual* was developed as a companion document to this *Final Report*. It provides a description of the variable fields for the 126 variables in the *Vehicle* data set and the 91 variables in the *Occupant* data set to be used for statistical analysis purposes. It also presents information in tabular format for each variable - the SAS Variable Name, Format Type, Field Length, Range of Values, Data Set (Vehicle or Occupant), Unique Identifier, and NHTSA Form Number (1002, 1002A, 1002B, 1002C).

Included in the *Coding Manual* are four appendices. Appendix A provides the specific codes assigned in this project to vehicle makes and models. Appendix B contains copies of the interview and observation survey forms that have been modified to include "unique identifiers", which were assigned to each interview question or observational element that is coded in the data tables, for ease of referencing between the survey forms and the data tables. Appendix C presents information concerning the data check and data entry procedures that were utilized, while Appendix D provides a glossary of child restraint terms.

---

[11] Copies of the data collection forms are presented in Appendix B of the *Training Manual*, at the end of the report.

SAS File

The Microsoft Access database used for data entry was converted to an SAS file with a *Vehicle* data set and an *Occupant* data set. The *Vehicle* data set contains information describing the vehicle make, model, year, and restraint technology for each seating position; characteristics of the site where data was collected; and responses to questions asked of drivers about their knowledge of LATCH technology and their knowledge of the different stages of child occupant protection. The *Occupant* data set contains occupant demographics and information about restraint use.

ANALYSIS OF SELECTED DATA

This chapter presents the results of selected data analyses in both narrative and tabular format. A general summary of the target drivers, vehicles, and young child occupants is provided first, for Survey Groups 1 through 4. An overall summary of data on LATCH system use for occupants less than 5 years old in a CRS in the back seat of a vehicle with upper tether anchor only (Survey Group 3), or upper tether anchor and lower anchors (Survey Group 4) is then presented.

Next, for the sample of upper tether users, detailed summaries are presented describing observed upper tether use—proper use, misuse, and nonuse. This is followed by summaries of interview data from nonusers of the upper tether strap, describing drivers' reasons for not attaching the upper tether strap. This section closes with ease-of-use ratings obtained from drivers of vehicles in which upper tethers were being used.

Analyses describing lower attachment use are presented following the discussion of upper tether use. For the sample of lower attachment users, detailed summaries are presented describing observed lower attachment use—proper use, misuse, and nonuse. This is followed by summaries of interview data from nonusers of the lower attachments, describing drivers' reasons for not attaching the lower attachments. The section on lower attachment use closes with ease-of-use ratings obtained from driver interviews with the lower attachment users, and drivers' preferences for attaching their CRSs to the vehicle (lower attachment versus the seat belt).

Attachment G presents general summary data on child occupants younger than 5 years old in booster seats; restraint use of all children age 5 to 13; and appropriateness of restraint selection for all child occupants younger than age 13.

TARGET SAMPLE CHARACTERISTICS

Observations were made between April and October 2005 for 1,182 drivers/vehicles and 1,728 child occupants younger than 13. The number of target vehicles observed in each State was as follows: Arizona, 163; Florida, 73; Michigan, 271; Missouri, 120; North Carolina, 171; Pennsylvania, 113; and Washington, 271. More than half of the data collection sites (55%) were shopping centers; 17 percent were child care centers; 12 percent were recreation facilities (e.g., swimming pool clubs); and the remainder were fast-food restaurants, health-care facilities, or other locations frequently visited by drivers with young children. Most of the sites were in either suburban (48%) or urban (46%) areas. In addition, 80 percent of the sites were in areas defined by the State site coordinators as middle- to upper-middle socio-economic class.

Seventy-two percent of the drivers were female. While a large percentage of the drivers were white (77%), 9 percent were African American, 7 percent were Latino/Hispanic, 5 percent were Asian, and the remainder were Pacific-Islander, Native American, or Other. Of the 1,182 drivers who participated in the interviews, 52 percent of them had heard of a new way to install a CRS without a vehicle seat belt. Of this subgroup of 609 drivers who had heard of a new way, 69 percent reported that it was called "LATCH."

Over 70 percent of the vehicles were model year 2000 and later; and over 50 percent of the vehicles were model year 2002 and later. Over half of the vehicles were either minivans or SUVs (52%); and 38 percent were 4-door passenger cars.

The main focus of this analysis is children less than 5 years old, as they are the intended users of child restraint systems in general, and LATCH technology in particular. This is because children 5 and older, depending upon weight and height, may be riding restrained on booster seats (using the vehicle seat belt as the restraint) or in CRSs designed for higher-weight children. Booster seats without an internal harness and higher-weight CRSs are not intended to be attached to vehicles using LATCH, as most LATCH anchors are rated for children who weigh less than 40 (and sometimes 48) pounds. Brief summaries of results pertaining to children younger than 5 who were riding in booster seats, and child occupants age 5 to 12 are presented in Attachment G at the end of this report.

There were 1,351 child occupants younger than 5 in total, in Groups 1 to 4. Table 1 presents the number of children by observed restraint type.

Table 1. Restraint Type for Child Occupants Younger Than 5 Years Old.

Restraint Type	Number (Percent) of Children
CRS	1,134 (84%)
Booster	195 (14%)
Lap And Shoulder Belt	15 (1%)
Lap Only Belt	3 (<1%)
Unrestrained	4 (<1%)
Total	1,351

All but 14 of the children under age 5 were riding in the back seat. The 14 children riding in the front seat were restrained as follows: 2 children were in rear-facing CRSs, 9 were in forward-facing CRSs, 2 were in belt-positioning boosters, and 1 was in a lap and shoulder belt. With regard to the 2 rear-facing children, one child was in a vehicle without an airbag switch and the other child was in a vehicle with an airbag switch in the "on" position. This child was riding in an infant-only CRS without a base, and the CRS was not attached to the vehicle.

The extremely low proportion of children who were: (1) unrestrained, and (2) riding in the front seat probably was associated with the following survey features: (1) the survey data collectors concentrated on vehicles with children restrained in the back seat of the vehicle, and (2) parents who agreed to participate in the survey were probably whose who restrained their children and placed them in the back seat.

The main focus of the study was to determine LATCH system use and misuse with child occupants younger than 5 years old in a CRS. As stated earlier, it is important to note that it was not required that the child be riding in a seating position equipped with lower anchors and/or a upper tether anchor; it only required that the child be riding in a vehicle equipped with lower anchors and/or upper tether anchors somewhere in the vehicle.

There were 602 Group 4 children younger than 5 being transported in a CRS with an upper tether strap and a lower attachment, which was located in the front or back seat of a vehicle that was equipped with both an upper tether anchor and lower anchors somewhere in the vehicle.

Upper Tether and Lower Attachments Use

Table 2 presents a comparison of all the methods used to attach the current CRSs to current vehicles. It is limited to:

- Vehicles equipped with the full LATCH system (upper tether and lower anchors);
- CRSs equipped with upper tethers and lower attachments;
- CRSs used in a mode (usually forward-facing) where upper tethers and lower attachments are recommended.

Not all seating positions in LATCH-equipped vehicles are themselves LATCH-equipped. It was important to determine whether drivers who were transporting children in LATCH-ready CRSs (i.e., those with lower attachments and upper tether straps) were taking advantage of lower and upper tether anchors in their vehicles, to attach the CRSs. The data examined for this analysis was limited to the subset of children riding in CRSs equipped with upper tethers and lower attachments and for which the use of an upper tether strap is appropriate. The first column presents the number and percent of CRSs attached using each method, when both lower anchors and upper tether anchors were available somewhere in the vehicle. The second column presents the number and percent of CRSs attached using each method, when both lower anchors and an upper tether anchor were available in the child's seating position in the vehicle. Overall, 208 of these 375 children, 55 percent, were using the upper tether. This is a huge improvement over the 15 percent use of upper tethers in safety seats of the 1980s, when vehicles were not required to have upper tether anchorages.[12] Forty-five percent of parents and caregivers, however, are still not using the upper tethers.

---

[12] Kahane, Charles J., 1986. An Evaluation of Child Passenger Safety: The Effectiveness and Benefits of Safety Seats. DOT HS 806 890. National Highway Traffic Safety Administration, Washington, DC.

Table 2. Observed CRS Attachment to the Vehicle, for Children Under 5, Riding in a CRS Equipped With <u>Both</u> an Upper Tether Strap and Lower Attachments for Which the Use of an <u>Upper Tether Strap Is Appropriate</u>, as a Function of LATCH Availability (in the vehicle somewhere versus in the child's particular seating position).

Method Used to Attach CRS to Vehicle When Use of Upper Tether Strap Is Appropriate	Number (Percent) for Children in Vehicles Equipped With Lower Anchors <u>and</u> Upper Tether Anchors Somewhere in the Vehicle (n=426)	Number (Percent) for Children in Vehicles Equipped With Lower Anchors <u>and</u> a Upper Tether Anchor in the Child's Particular Seating Position (n=375)
Upper Tether and Lower Anchors	149 (35%)	138 (37%)
Upper Tether and Lower Anchors and Seat Belt	29 (7%)	26 (7%)
Upper Tether and Seat Belt	59 (14%)	44 (12%)
Upper Tether Only	0	0
Lower Anchors and Seat Belt	20 (5%)	20 (5%)
Lower Anchors Only	37 (9%)	37 (10%)
Seat Belt Only	132 (31%)	110 (29%)
Not Attached	0	0

Table 3 present a comparison of methods used to attach the CRSs to vehicles. It is limited to:
- Vehicles equipped with the full LATCH system (upper tether and lower anchors);
- CRSs equipped with upper tethers and lower attachments;
- CRSs used in a mode (usually rear-facing) where only the lower attachments are recommended.

Since upper tether use is not considered appropriate for these CRSs by the manufacturers, the 6 percent of CRSs attached to the vehicle using an upper tether (i.e., "Tether and Lower Anchors and Seat Belt" or "Tether and Seat Belt") are not attached in a recommended way. As in Table 2, some CRSs were attached with seat belts instead of lower attachments and others with seat belts in addition to lower attachments.

Table 3. Observed CRS Attachment to the Vehicle, for Children Under 5, Riding in a CRS Equipped With <u>Both</u> an Upper Tether Strap and Lower Attachments for Which the Use of an <u>Upper tether Strap Is Not Appropriate</u>,[13] as a Function of LATCH Availability (in the vehicle somewhere versus in the child's particular seating position).

Method Used to Attach CRS to Vehicle When Use of Upper Tether Strap Is Not Appropriate	Number (Percent) for Children in Vehicles Equipped With Lower Anchors <u>and</u> Upper Tether Anchors Somewhere in the Vehicle (n=85)	Number (Percent) for Children in Vehicles Equipped With Lower Anchors <u>and</u> an Upper Tether Anchor in the Child's Particular Seating Position (n=69)
Upper Tether and Lower Anchors and Seat Belt	2 (2%)	2 (3%)
Upper Tether and Seat Belt	2 (2%)	2 (3%)
Lower Anchors and Seat Belt	7 (8%)	6 (9%)
Lower Anchors Only	44 (52%)	38 (55%)
Seat Belt Only	30 (35%)	21 (30%)
Not Attached	0	0

Table 4 combines the observations in Table 2 and Table 3, but focuses only on the use of lower attachments and/or seat belts.

Table 4. Observed CRS Attachment to the Vehicle, for Children Under 5, Riding in a CRS Equipped With <u>Both</u> an Upper Tether Strap and Lower Attachments, as a Function of LATCH Availability (in the vehicle somewhere versus in the child's particular seating position).

Method Used to Attach CRS to Vehicle, Grouped by Type of Method	Number (Percent) for Children in Vehicles Equipped With Lower Anchors <u>and</u> Upper Tether Anchors Somewhere in the Vehicle (n=511)	Number (Percent) for Children in Vehicles Equipped With Lower Anchors <u>and</u> an Upper Tether Anchor in the Child's Particular Seating Position (n=444)
Lower Attachments	230 (45%)	213 (48%)
Lower Attachments & Seat Belts	58 (11%)	54 (12%)
Seat Belts	223 (44%)	177 (40%)
Neither	0	0

Of all 511 children (426 from Table 2 and 85 from Table 3) in LATCH-equipped vehicles, 67 (13 percent) are not riding at LATCH-equipped positions. Many of these children are riding in vehicles equipped with LATCH at the rear-outboard seats but not the rear-center seat. The parents and caregivers are placing the safety seat in the center-rear seat of the vehicle.

---

[13] The manufacturers of the CRSs do not recommend the use of the tether for these CRSs. They are mostly rear-facing CRSs.

which is generally considered the safest position[14], instead of placing it in one of the LATCH-equipped rear-outboard seats.

Of the 444 children in LATCH-equipped seating positions, 48 percent of parents and caregivers are attaching the CRS to the vehicle using lower attachments (LATCH in child's seating position); an additional 12 percent are using both lower attachments and seat belts. Forty percent of parents and caregivers are using seat belts, or seat belts and upper tether, without using lower attachments.

Thus, with the latest-model vehicles and CRSs equipped with the full LATCH systems:

- 55 percent of upper tethers are actively in use, for configurations where use of an upper tether is recommended.
- 60 percent of parents and caregivers are using lower anchors to install the safety seat.

Correct Use and Misuse of Upper Tethers

Data describing upper tether use for children younger than 5 are presented next. For the 317 young child occupants in the back seat in CRSs with an upper tether in use, 13 were in rear-facing CRSs, 301 were in forward-facing CRSs with a harness, and 3 were in combination seats with both the internal harness and the vehicle seat belt restraining the child. The CRS upper tether strap adjustor type on these 317 seats was as follows: 75 percent were tilt-lock, 15 percent were double-back; and 15 percent were some other type. (See Attachment A for definitions of the different types of adjustors used in upper tether straps and lower attachments.)

Many criteria determine proper upper tether use. Table 5 presents proper-use data for each proper-use criterion observed, for the 304 forward-facing CRSs (301 forward-facing with harness and 3 combination seats).

---

[14] Kahane, Charles J., Lives Saved by the Federal Motor Vehicle Safety Standards and Other Vehicle Safety Technologies, 1960-2002 – Passenger Cars and Light Trucks – With a Review of 19 FMVSS and Their Effectiveness in Reducing Fatalities, Injuries and Crashes. NHTSA Technical Report Number HS 809 833. National Highway Traffic Safety Administration, Washington, D.C., October 2004. p. 134.

Table 5. Observations of Proper Upper Tether Use for 304 Forward-Facing CRSs.

Proper Upper Tether Use Criterion	Number (%) of CRSs (n=304)
One Upper Tether per Upper Tether Anchor	294/295 (99.7%)
Upper Tether Tightness Allows CRS Base to Rest Flat on Vehicle Seat	291/295 (99%)
Upper Tether Is Attached to Upper Tether Anchor Designated for Child's Seating Position	271/281 (96%)
Upper Tether Is Routed Properly (e.g.. over an integrated vehicle head restraint or under an adjustable vehicle seat head restraint).	266/299 (89%)
Upper Tether Is Flat (Flat or twisted no more than ½ twist)	261/297 (88%)
Upper Tether Is Snug (Passes "pinch test")	244/298 (82%)

Note: Denominator may be less than 304. due to missing data.

Upper tether <u>misuse</u> characteristics for the 304 forward-facing CRSs with upper tether straps in use are shown in Table 6.

Table 6. Observed Upper Tether Installation Misuse Characteristics for the 304 Forward-Facing CRSs Equipped With an Upper Tether Strap.

Upper Tether Misuse[15]	Number (%) of Forward-Facing Seats (n=304)
Upper Tether Is Loose	54/298 (18%)
Upper Tether Is Twisted	36/297 (12%)
Improper Routing: Over a Down-Adjustable Head Restraint	21/299 (7%)
Improper Routing: Over a Raised Adjustable Head Restraint	6/299 (2%)
Improper Routing: Wrapped Around Read Restraint	2/297 (<1%)
Attached to Something Other Than to a Designated Upper Tether Anchor	10/281 (4%)
Upper Tether Used on Combination Seat in Booster Mode	3/295 (1%)
Upper Tether Is Too Tight (Pulls CRS Base off Vehicle Seat)	4/295 (1%)
More Than 1 Upper Tether Attached Per Upper Tether Anchor	1/295 (<1%)

Note: Denominators may be less than 304. due to missing data.

Nonuse of Upper Tethers

For the 398 CRSs with upper tether straps in vehicles with upper tether anchors, where the upper tether strap was not in use (292 forward-facing CRSs and 106 rear-facing CRSs), 82 percent were stowed and 18 percent were hanging loose.[16]

When drivers were asked why the upper tether was not being used. the reasons shown in Table 7 were provided by the drivers for 207 of the forward-facing CRSs. Multiple answers

---

[15] Other tether use that was observed included: 12/297 (4%) CRSs were used for children who weighed 40 or more pounds.

[16] Upper tether straps that are hanging loose can become tangled around or hit a child or another passenger in the vehicle. Therefore. it is preferred that upper tether straps are stowed when not in use.

were allowed. (Interviewers circled as many unprompted responses as were given, with non-standard responses recorded in the additional space provided on the form for "Other") Sixty-one percent of the upper tether nonusers provided at least one of the responses related to a lack of knowledge about upper tethers (e.g., about their existence, their importance, how to use them) as their reason for not using them.

Table 7. Reasons Provided for Nonuse of Upper Tether Anchors, for Upper Tether-Equipped CRSs Installed in Vehicles With Upper Tether Anchors (somewhere in the vehicle).

Reason For Nonuse Of Upper Tether Anchors	Number (%) of CRSs
"I didn't know how to use it"	75/207 (36%)
"I didn't think it was important to use"	52/207 (25%)
"Vehicle does not have equipment/anchor/place to hook"	21/207 (10%)
"I didn't know about it"	10/207 (5%)
"Too hard to use"	11/207 (5%)
"Switch CRS back and forth between cars – forget to hook"	10/207 (5%)
"Don't know where it goes"	4/207 (2%)
"Switch CRS back and forth between cars – too much trouble"	4/207 (2%)
"Seating position does not have equipment/anchor/place to hook"	3/207 (1%)
"Too much trouble/inconvenient"	3/207 (1%)
"Brand new car"	2/207 (1%)
"Can't get tight enough"	2/207 (1%)
"Use seat belt instead"	2/207 (1%)
"Switched CRS from rear-facing to forward-facing and forgot to attach"	1/207 (<1%)
"Complicates things/Just as tight without it"	1/207 (<1%)
"Don't see the point"	1/207 (<1%)
"Husband installed seat"	1/207 (<1%)
"Position of seat in car and number of occupants"	1/207 (<1%)
"No reason"	1/207 (<1%)
"Inconvenient with 3rd row seat"	1/207 (<1%)
"Not our car"	1/207 (<1%)
"Can get seat belt tighter"	1/207 (<1%)
"It runs through the back of the chair"	1/207 (<1%)
"Used to (use it)"	1/207 (<1%)
"Car doesn't need it"	1/207 (<1%)

Note: Of the 398 CRSs with upper tether straps, 106 were rear-facing CRSs; the drivers of the vehicles with these CRSs were not asked why they did not use the upper tether anchors. Of the 292 forward-facing CRSs, 85 drivers did not provide reasons for nonuse of upper tether anchors.

With regard to the 21 responses that the vehicle doesn't have the equipment, it should be noted that all vehicles in the analysis for Table 7 had upper tether anchors <u>somewhere in the vehicle</u>. Regarding the 3 responses that the seating position did not have the equipment, 2 of the CRSs were a center-rear seat that was equipped with an upper tether anchor, and 1 CRS was in a third-row seat that did not have an upper tether anchor.

The seating positions of the 292 forward-facing, tether-equipped CRSs (for which the upper tether was not in use) in vehicles where an upper tether anchor was present somewhere in the vehicle, and the percentage of vehicles where an upper tether anchor was available <u>for that seating position</u> are shown in Table 8. It shows that 97 percent of the CRSs with tethers not in use had been placed in seating positions that were equipped with upper tether anchors.

Table 8. Seating Positions and Percentage of Vehicles With an Upper Tether Anchor in the Seating Position, for 292 Forward-Facing, Upper Tether-Equipped CRSs Where the Upper Tether Was Not in Use.

Seating Position	Number (%) of CRSs in This Position	Number (%) of Vehicles With a Upper Tether Anchor in This Position
Second Row Left	84 (29%)	84 (100%)
Second Row Center	47 (16%)	44 (94%)
Second Row Right	151 (52%)	150 (99%)
Third Row Left	4 (1%)	2 (50%)
Third Row Center	2 (<1%)	2 (100%)
Third Row Right	4 (1%)	1 (25%)
Total	292	283 (97%)

Upper Tether Ease-of-Use Ratings

Table 9 provides ease-of-use ratings that were obtained from the drivers for 281 of the 305 CRSs being used with an upper tether. For the forward-facing CRSs, upper tether use was rated as "very easy" or "relatively easy" for 81 percent, "very difficult" or "somewhat difficult" for 13 percent, and neither easy nor difficult for 6 percent of the CRSs. In addition, drivers rated the ease of use of the tilt-lock adjuster and the double-back adjuster. Both were rated by a large majority of drivers as being either very easy or relatively easy to use; however, a higher percentage of drivers found the tilt-lock adjuster easy to use than did the drivers using CRSs with the double-back adjuster on their upper tether strap.

Table 9. Ease-of-Use Ratings for Drivers Using Upper Tethers to Install Forward-Facing CRSs in a Back Seating Position (for children under 5).

Category	Very Easy	Relatively Easy	Somewhat Difficult	Very Difficult	Neither Easy nor Difficult
Overall (n=305)					
Upper Tether (Forward-Facing CRSs) (n-305)	34% (96/281)	47% (131/281)	9% (24/281)	4% (12/281)	6% (18/281)
Upper Tether Strap Adjuster Type					
Tilt-Lock (n=229)	33% (71/216)	51% (110/216)	6% (13/216)	3% (6/216)	7% (16/216)
Double-Back (n=46)	37% (15/41)	32% (13/41)	20% (8/41)	10% (4/41)	2% (1/41)

Note: Denominator may be less, due to missing data.

Correct Use and Misuse of Lower Attachments

Data describing lower anchor use are presented next, for child occupants younger than 5. There were 353 young child occupants in back seats in CRSs that were using the lower attachments. One of these children was in a CRS using lower attachments in a vehicle that had no lower anchors; this child is coded as Group 3 (because the vehicle has upper tether anchors but no lower anchors), whereas the other 352 children are coded as Group 4. Regarding CRS lower attachment connector types for the 353 CRSs, 82 percent were flexible strap, hook-on; 15 percent were flexible strap, push-on; and 1 percent were rigid attachments. Regarding webbing tension release characteristics, 44 percent were squeeze-release, 49 percent were tilt-lock, and 5 percent were unknown. For strap type on the CRS, 65 percent were single straps; 8 percent were side straps; 4 percent had 2 straps, and 23 percent were "unsure". (See Attachment A for definitions.)

Many criteria determine proper lower anchor use. Table 10 presents proper-use data for each proper-use criterion observed, for the 353 CRSs attached to the vehicle using lower attachments. Data for rear-facing and forward-facing CRSs are combined, as both types were designed to be attached using lower anchors.

Table 10. Observations of Proper Lower Attachment Use for 353 CRSs.

Proper Lower Attachment Use Criterion	Number (%) of CRSs
One Connector per Bar	344/349 (99%)
Correct Path Used for Flexible Single Strap	282/303 (93%)
Connectors Installed Right Side Up	316/346 (91%)
Lower Attachment Strap Is Flat (flat or twisted no more than ½ twist)	310/339 (91%)
CRS Attached to Lower Anchors Designated for Child's Seating Position	308/348 (89%)
Correct Angle Is Achieved for Rear-Facing CRSs	93/117 (79%)
Installation Is Tight (passes 1-inch test)	246/349 (70%)

Note: Denominator may be less than 353, due to missing data, or due to type of attachment or type of CRS.

Sixty-one percent of seats installed with lower attachments were securely installed—the connectors were installed right side up, the lower attachment straps were flat and routed to the correct anchors, and the installation was tight. In the last survey before LATCH, less than 50 percent of seats installed with seat belts were securely installed (i.e., attachment by the vehicle seat belt was not loose).[17]

---

[17] Decina, L.E., and Lococo, K.H., 2004. Misuse of child restraints. DOT HS 809 671. U.S. Department of Transportation/National Highway Traffic Safety Administration, Washington, DC.

Lower anchor <u>misuse</u> characteristics for the 353 CRSs being used with lower attachments by children under age 5 are shown in Table 11.

Table 11. Observations of Lower Anchor Misuse, for 353 CRSs Installed Using the Lower Attachments.

Lower Anchor Misuse[18]	Number (%) of CRSs
Loose Installation (e.g., can move CRS more than 1 inch from side-to-side)	103/349 (30%)
Lower Attachment Strap Twisted	29/339 (9%)
Lower Connector Installed Up-Side Down	30/346 (9%)
CRS Attached to Something Other Than Designated Lower Anchors	40/348 (11%)
Using Lower Anchors Not Designated for Seating Position	27/348 (8%)
In Lower-Anchor Designated Seating Position, But Only One Side Attached to Anchor	6/348 (2%)
In Lower-Anchor Designated Seating Position, But Attached to Something Other Than Anchor (e.g., Spring or Seating Material)	2/348 (1%)
Lower Attachments Hooked Through Bight of Seat to Each Other	1/348 (<1%)
Incorrect Angle Achieved for Rear-Facing Seats	24/117 Rear-facing seats (21%)
Combination CRS Used in BPB Mode	8/123 Combo Seats (7%)

Note: Denominator may be less than 353, due to missing data, or due to type of attachment or type of CRS.

Of the 27 CRSs identified as using lower anchors not designated for the child's seating position, 26 (96 percent) were in a center-rear seat.

Drivers who attached their CRSs to the vehicle using both the lower attachments and the vehicle seat belt (observed for 69 of the 353 lower attachment users) were asked why they used both methods. The following responses were obtained: for extra secureness or safety (77%), they thought it was necessary (11%), and other reasons (13%). Other reasons included: they didn't personally install the CRS, to keep the CRS from moving or to get it tight, and "in case one breaks."

Overall Misuse for Secure Attachment to Vehicle

In the past 10 years, there have been a number of surveys that have measured the rate of misuse of attaching CRSs to vehicles, including the 1995 Ketron study[19] and the 2002

---

[18] Other types of lower anchor use that were observed included: 69/353 (20%) CRSs were installed using both the lower attachments and seat belts; and 9/353 (3%) CRSs were used for children who weighed 40 or more pounds (Note: none of the children weighed 48+ pounds).

[19] Decina, L.E., and Knoebel, K.Y., *Patterns of Misuse of Child Safety Seats*, NHTSA, Publication No. DOT HS 808 440, National Highway Traffic Safety Administration, Washington, DC, 1996; Decina, L.E., and Knoebel, K.Y., "Child Safety Seat Misuse Patterns in Four States," *Accident Analysis and Prevention*, Vol. 29, January 1997, pp. 125-132.

TransAnalytics study.[20] Comparing the misuse rates in these studies is difficult because of the type of child safety seats observed and the categories of misuse measured vary from study to study. Also the misuse rates tend to be counts of the different types of misuse, rather than the number of CRSs having at least one type of misuse. Therefore, the misuse rates often sum to be greater than 100%, due to some of the CRSs having more than one type of misuse. Within each study, however, types of misuse, as identified by certain grouping rules, can be combined together to produce an overall measure of the misuse related to the insecure attachment of the CRS to the vehicle. If misuse for boosters is excluded, as well as misuse which is related to placing the child into the CRS's harness system, the main remaining types of misuse are related to insecurely attaching the CRS to the vehicle with the seat belt (i.e., loose seat belt; misuse/nonuse of the locking clip) or with the lower attachments (i.e., loose, misrouted, twisted, or up-side down lower attachments).

As it can be seen in Table 12 below, the rate of misuse (for not tightly attaching the CRS to the vehicle) has been decreasing. There are two main reasons that misuse has decreased.

1. LATCH does not use locking clips. Instead it uses automatic or manual adjustors that are not prone to the type of misuse seen with locking clips. While the percent of CRSs with observed misuse for nonuse/misuse of the locking clip is 72 percent in the 1995 Ketron study and 4 percent in the 2002 TransAnalytics study, there is not any misuse associated with nonuse/misuse of the locking clip in the 2005 LATCH survey.

2. LATCH has a simpler installation, in comparison with the use of a seat belt. Improper or loose fitting seat belts were observed in 14 percent of the CRSs in the Ketron Study and 54 percent in the 2005 TransAnalytics study. The overall misuse observed in the 2005 LATCH survey was 39 percent when the survey data was analyzed to include only the number of CRSs with at least one of these types of misuse, rather than to count how many different types of misuse were found.[21]

In the "Overall Misuse" category below, a range is provided for the 1995 Ketron Study and for the TransAnalytics Study from the minimum to the maximum amount of misuse possible given the numbers in the "CRSs with Misuse (%)" column and given that the categories of misuse observed could overlap entirely or not at all. For the 2005 LATCH Survey, a percent (39%), not a range, is given because the overall misuse percent can be calculated from the survey data.

---

[20] Decina, L.E., and Lococo, K., *Misuse of Child Restraints,* NHTSA Publication No. DOT HS 809 671, National Highway Traffic Safety Administration, Washington, DC, 2004; Decina, L.E., and Lococo, K., "Child Restraint System Use and Misuse in Six States," *Accident Analysis and Prevention, Vol.* 37, 2005, pp. 583-590.
[21] While the percent of CRSs with loose (30%), misrouted (13%), twisted (9%) or up-side down (9%) attachments sums up to 61% in the 2005 LATCH survey (See Table 11), only 39% of the CRSs were observed to have one or more of these misuses.

Table 12. Observed Misuse Of Attaching CRSs To Vehicles.

Study	Type of Misuse Observed	CRSs with Misuse (%)	Overall Misuse (%)
1995 Ketron	Nonuse/Misuse of Locking Clip	72	
	Misrouted SB or Improper Seat Belt Fit	17	72-89
2002 TransAnalytics	Loose Seat Belt	54	
	Misrouted SB or Misuse of Locking Clip	6	54-60
	Loose Attachment	30	
	Misrouted Attachment	13	
	Twisted Attachment	9	
2005 LATCH Survey	Up-Side Down Attachment	9	39

Nonuse of Lower Attachments

For children in CRSs with lower attachments, in vehicles equipped with lower anchors, but the lower attachments were not in use (n=249), 80 percent were stowed, and 20 percent were hanging loose. The seating position and percentage of vehicle seats with lower anchors in the position occupied by the child is shown in Table 13. It shows that 26 percent of CRSs with lower attachments not in use had been placed on seats not equipped with lower anchors: the second-row center seat or a third-row seat.

Table 13. Seating Positions and Percentage of Vehicles With Lower Anchors in the Seating Position, for 249 Lower-Attachment-Equipped CRSs Where the Lower Attachments Were Not in Use.

Seating Position	Number (%) of CRSs in This Position	Number (%) of Vehicles With Lower Anchors in This Position
Second Row Left	71 (29%)	71 (100%)
Second Row Center	59 (24%)	15 (25%)
Second Row Right	104 (42%)	104 (100%)
Third Row Left	5 (2%)	3 (60%)
Third Row Center	5 (2%)	2 (40%)
Third Row Right	5 (2%)	1 (20%)
Total	249	196 (79%)

Drivers in vehicles with lower anchors who were transporting children in CRSs equipped with lower attachments but were not using them were asked why they were not using the lower attachments. The reasons provided by drivers for 145 of the CRSs are shown in Table 14. Multiple answers were allowed. (Interviewers circled as many unprompted responses as were given, with non-standard responses recorded in the additional space provided on the form for "Other".) Fifty-five percent of the lower attachment nonusers provided at least one of the responses related to a lack of knowledge about lower attachments (e.g., about their existence, their importance, how to use them) as their reason for not using the lower attachments.[22]

Table 14. Reasons Provided for Nonuse of Lower Anchors, for Lower-Attachment-Equipped CRSs Installed in Vehicles With Lower Anchors (somewhere in the vehicle).

Reason For Nonuse Of Lower Anchors	Number (%) of CRSs
"I didn't know how to use it"	54/145 (37%)
"I didn't think it was important to use"	24/145 (17%)
"No anchors" or "Vehicle isn't equipped"	17/145 (12%)
"No anchors in this seat position"	12/145 (8%)
"Too hard to use"	11/144 (8%)
"I didn't know about it"	7/145 (5%)
"Can't get seat installed tightly"	5/140 (4%)
"Using seat belt"	3/145 (2%)
"More secure with seat belt"	3/145 (2%)
"Husband installed CRS"	2/145 (1%)
"It's a new car"	2/145 (1%)
"Middle is the safest position"	1/145 (<1%)
"Seat belt is easier"	1/145 (<1%)
"Don't need"	1/145 (<1%)
"Don't like"	1/145 (<1%)
"Just not using"	1/145 (<1%)
"Hard switching to another vehicle"	1/145 (<1%)
"Switch car"	1/145 (<1%)
"Didn't read instructions"	1/145 (<1%)
"Seat belt has locking mechanism"	1/145 (<1%)
"Know how to use seat belt"	1/145 (<1%)
"Mom took out to wash and forgot to reinstall"	1/145 (<1%)
"Rear-facing seat"	1/145 (<1%)

Fifty-four drivers provided reasons other than the 5 listed on the data collection form for this question. Two categories of interest emerged when these reasons were grouped together.

- Back-Center Seating Position. Drivers stated that lower anchors were not available in this seating position or that they believed that the middle position was safer.

- Seat Belts. Drivers stated that they thought seat belts were safer or better than lower attachments or that they used seat belts because they knew how to use them.

---

[22] In response to the question. "If NO, why aren't you using it?" the driver provided one or more of the following answers: (1) Didn't know about it. (2) Did not think it was important to use. (3) Don't know how to use it.

The seating positions of the 54 CRSs for which "other" reasons were provided for nonuse of the lower attachments and the percentage of vehicles with lower anchors in those positions are shown in Table 15.

Table 15. Seating Positions of the 54 Children For Whom "Other" Reasons for Nonuse of the Lower Anchors Were Offered, and Percentage of Vehicles Equipped With Lower Anchors in Those Seating Positions.

Seating Position	Number (%) of CRSs in This Position	Number (%) of Vehicles With Lower Anchors in This Position
Second Row Left	11 (20%)	11 (100%)
Second Row Center	23 (43%)	7 (30%)
Second Row Right	14 (26%)	14 (100%)
Third Row Left	2 (4%)	0
Third Row Center	2 (4%)	0
Third Row Right	2 (4%)	0
Total	54	32 (59%)

Of particular interest are the 29 responses that indicated that either "the vehicle was not equipped with lower anchors" or the "seating position was not equipped with lower anchors", even though all of the vehicles included in the analyses reported in Tables 9 and 10 were equipped with lower anchors <u>somewhere in the vehicle</u>. This response was provided for 10 CRSs in the second row outboard position; all 10 of these vehicles were in fact equipped with lower anchors in the positions occupied by the children. This response was offered for 14 CRSs in the second row center; only 2 of the 14 vehicles were equipped with lower anchors in the second row center seat. This response was offered for 5 CRSs in the third row of vehicle seats. None of the 5 vehicles were equipped with lower anchors in the seating positions occupied by the children.

Lower Attachment Ease-of-Use Ratings

Table 16 provides ease-of-use ratings obtained from drivers who had connected their CRSs to the vehicle using the lower attachments. Details are provided on overall ease of use and on ease of use by the type of hardware used with the lower attachment (for all CRSs). Of the 328 CRSs for which ratings were obtained, 74 percent indicated that using the hardware was either "very easy" or "relatively easy." Eighteen percent of the CRS connections were rated as either "somewhat difficult" or "very difficult". There is insufficient data to make meaningful comparisons among the three different types of connectors for lower attachments (i.e., rigid, flexible strap with push-ons or with hook-ons), but drivers using CRSs with the webbing tension release hardware rated both types (tilt-lock and squeeze-release) about the same.

**Table 16. Ease-of-Use Ratings for Drivers Using Lower Attachments to
Install Forward-Facing and Rear-Facing CRSs in a Back Seating Position
(for children under 5).**

Category	Very Easy	Relatively Easy	Somewhat Difficult	Very Difficult	Neither Easy nor Difficult
Overall					
Lower Attachments (All CRSs) (n=354)	39% (127/328)	35% (115/328)	13% (44/328)	4% (14/328)	9% (28/328)
Lower Attachments--Connector Type					
Rigid (n=5)	80% (4/5)	20% (1/5)	0% (0/5)	0% (0/5)	0% (0/5)
Flexible Strap w/Push-Ons (n=53)	42% (21/50)	34% (17/50)	10% (5/50)	2% (1/50)	12% (6/50)
Flexible Strap w/Hook-Ons (n=288)	38% (100/266)	36% (96/266)	14% (37/266)	5% (12/266)	8% (21/266)
Lower Attachments--Webbing Tension Release Type					
Tilt-lock Users (n=173)	43% (69/160)	32% (51/160)	10% (16/160)	5% (8/160)	10% (16/160)
Squeeze-release Users (n=155)	32% (47/145)	39% (56/145)	19% (27/145)	3% (5/145)	7% (10/145)

Note: Denominator may be less, due to missing data.

Drivers who were using the lower attachments to connect their CRSs to the vehicle, and indicated that they personally installed the CRS (246 drivers) were asked what they like about LATCH. Multiple responses were permitted. Ten of the 246 respondents (4%) indicated that the question was not applicable (i.e., because they didn't like LATCH). The responses obtained are shown in Table 17.

Table 17. Responses Obtained by 246 Lower Attachment Users (who personally installed their CRSs) Describing What They Like About LATCH.

What Drivers Like About LATCH	Number (%) of Responses
"It's easy to use"	158/246 (64%)
"Results in a tight fit for the child safety seat"	74/246 (30%)
"Safer/More Secure"	22/246 (9%)
"Can see the connectors"	11/246 (5%)
"Don't know"	3/246 (1%)
"Added Safety" (they were using seat belt & lower anchors)	2/246 (<1%)
"Both car and CRS have it"	2/246 (<1%)
"Doesn't cause folding seat belts"	2/246 (<1%)
"Everything"	2/246 (<1%)
"Easy to remove"	1/246 (<1%)
"Easy to figure out"	1/246 (<1%)
"Easier for other kids to get in and out of the van"	1/246 (<1%)
"Only one I know"	1/246 (<1%)
"It works"	1/246 (<1%)
"It worked well in an accident"	1/246 (<1%)
"Tighten reinforcement"	1/246 (<1%)
"Quicker"	1/246 (<1%)
"Attach to vehicle"	1/246 (<1%)
No Response	1/246 (<1%)

These 246 drivers were also asked what they don't like about LATCH. Multiple responses were permitted. There were 102 respondents (42%) who indicated that the question was not applicable (i.e., they don't dislike anything about LATCH). The responses obtained are shown in Table 18.

Table 18. Responses Obtained by 246 Lower Attachment Users (who personally installed their CRSs) Describing What They Don't Like About LATCH.

What Drivers Don't Like About LATCH	Number (%) of Responses
"It's hard to release the CRS from the bars"	67/246 (27%)
"It's hard to hook the CRS to the bars"	29/246 (12%)
"It's hard to use"	22/246 (9%)
"I can't get the CRS tight"	20/246 (8%)
"It's hard to see the bars"	14/246 (6%)
"It's hard to find the bars"	6/246 (2%)
"It's not on all cars"	3/246 (1%)
"It's hard to use the first time"	3/246 (1%)
"The tether is hard to tighten"	2/246 (<1%)
"The instructions are unclear"	2/246 (<1%)
"It's hard to find when traveling"	2/246 (<1%)
"Tether angle"	1/246 (<1%)
"So much work, especially if you have to move the seat"	1/246 (<1%)
"Need to release the seat and bring it forward"	1/246 (<1%)
"Need to put muscle into it"	1/246 (<1%)
"In sedans, the bars are off-center"	1/246 (<1%)
"It's hard to tighten up"	1/246 (<1%)
"It's hard to get fingers in"	1/246 (<1%)
"It's hard to position self to put weight on seat"	1/246 (<1%)
"Climbing in to ensure tightness"	1/246 (<1%)
"Angle at seat difficult to connect LATCH"	1/246 (<1%)

Preferred Method for Attaching CRS

Drivers who had installed their CRSs using the lower anchors, but reported that they also had experience attaching a CRS to a vehicle using the vehicle seat belt (n=216) were asked which method they preferred. Seventy-five percent with experience using both methods reported a preference for LATCH, 9 percent preferred seat belts, and 16 percent were undecided. Reasons for preferring LATCH included: easier, safer, more secure, and tighter. Drivers who preferred the seat belt to LATCH indicated that they knew what to do with the seat belt, it was easier and quicker, and they couldn't get the CRS tight enough without the seat belt. When asked whether it was easier to install a CRS using the lower anchors or the seat belt, 69 percent reported lower anchors, 17 percent reported seat belts, and 15 percent were undecided.

DISCUSSION OF FINDINGS

This field observation study provided a look into LATCH system use and misuse among the motoring public, identified the public's knowledge of LATCH systems, and surveyed drivers' opinions of the ease of use of upper tethers and lower attachments. The first and foremost finding was the need for additional consumer education in several areas, such as:

- The survey showed that many parents were unaware of upper tethers' existence or their importance. Three-fifths of the upper tether nonusers provided at least one of the responses related to a lack of knowledge about upper tethers as their reason for not using them.

- When parents and caregivers had experience attaching CRSs using the seat belt and using lower anchors, three-fourths reported a preference for LATCH, because they found LATCH was easier to use and provided a tighter fit. Some parents and caregivers who currently are using seat belts to secure the CRS may feel more confident using the seat belt because they are familiar with how to secure a CRS this way, and they may not feel comfortable in learning to secure the CRS in a new way.

- Data collectors found a wide variety in accessibility of LATCH hardware in vehicles. For example, accessibility of the upper tether anchor varied by type of vehicle and manufacturer. In SUVs and minivans, accessing the upper tether anchor could be as simple as standing at the vehicle door and inserting a hand between the seat back and the side of the door frame, yet as challenging as having to be in the vehicle at that seating position and reaching over the back of the seat or climbing into the third row and crouching over to find the anchor on the floor. Pickup trucks offer challenges as well. In many cases, the seat had to be folded forward to fasten the upper tether and then the upper tether straps had to be adjusted. Lower anchor accessibility could also be quite challenging. In many cases, the lower anchor bar was embedded quite deeply into the seat bight, making it difficult to reach and connect the lower attachment.

- The center-rear seat is the safest position in a vehicle. But this seating position is not equipped with lower attachments in many passenger cars. Many parents are aware that the center seat is the safest position, so much of the nonuse of lower attachments in this study related to the fact that the vehicle seat belt was the only method available in that position for installing a CRS. In addition, many of the LATCH misuse issues observed in this study were related to this seating position. Some of the parents who chose to install their CRS using the lower anchors, positioned their child in the center-rear seat, which was not lower-anchor equipped, and then attached the CRS to the anchors designated for the outboard seating positions.

ACKNOWLEDGEMENTS

The authors wish to thank many people and organizations for their time and effort. First, they would first like to express their appreciation to the State site coordinators (SSCs) who were responsible for providing expert opinion on many facets of the project, as well as supervising and managing field operations in their respective States. They are: Arizona - Nancy Avery (Tucson Safe Kids), Florida - Karen Hanawalt (Tallahassee Community College), Michigan - Janelle Rose (Program Professionals, Inc.), Missouri - Cathy Metzger Hogan (Safe Kids St. Louis), North Carolina - William Hall (Highway Safety Research Center, University of North Carolina), Leigha Shepler (Safe Kids Guilford County), Amy Krise (Safe Kids Charlotte Mecklenburg), Pennsylvania - Juli McGreevy, Sherri Penchishen (Bethlehem Health Department), Cynthia Cianciulli (Montgomery County Highway Safety Program), and Washington - Kathy Kruger (Washington Safety Restraint Coalition).

Thanks are also in order to Deborah K. Stewart (Safe Ride News), Lorrie Walker (Safe Kids Worldwide), and Capt. Fred Mills (Bethlehem Police Department) for their contributions and valuable insights throughout the project.

Final thanks to the dozens of certified child passenger safety technicians and instructors who were responsible for collecting quality child occupant protection data in the seven States. Also thanks to Carol Martell of the University of North Carolina Highway Safety Research Center for converting our database to SAS-readable data files, and performing initial SAS runs.

REFERENCES AND BIBLIOGRAPHY

Code of Federal Regulations, Title 49. Government Printing Office, Washington, DC, 2005.

Decina, L.E. and Knoebel, K.Y. *Patterns of Misuse of Child Safety Seats.* NHTSA. Publication No. DOT HS 808 440, National Highway Traffic Safety Administration, Washington, DC, 1996.

Decina, L.E., and Knoebel, K.Y. "Child Safety Seat Misuse Patterns in Four States," *Accident Analysis and Prevention*, Vol. 29, January 1997, pp. 125-132.

Decina, L.E., and Lococo, K. *Misuse of Child Restraints.* NHTSA Publication No. DOT HS 809 671, National Highway Traffic Safety Administration, Washington, DC, 2004.

Decina, L.E., and Lococo, K. "Child Restraint System Use and Misuse in Six States," *Accident Analysis and Prevention, Vol.* 37, 2005, pp. 583-590.

Decina, L.E., Lococo, K., and Block, A. *Misuse of Child Restraints: Results of a Workshop to Review Field Data Results.* NHTSA Traffic Safety Facts – Research Note No. DOT HS 809 851, National Highway Traffic Safety Administration, Washington, DC, March 2005.

Kahane, Charles J. *An Evaluation of Child Passenger Safety: The Effectiveness and Benefits of Safety Seats.* NHTSA Technical Report No. DOT HS 806 890, National Highway Traffic Safety Administration, Washington, DC, 1986.

Kahane, Charles J. *Lives Saved by the Federal Motor Vehicle Safety Standards and Other Vehicle Safety Technologies, 1960-2002 – Passenger Cars and Light Trucks – With a Review of 19 FMVSS and their Effectiveness in Reducing Fatalities, Injuries and Crashes.* NHTSA Technical Report No. DOT HS 809 833, National Highway Traffic Safety Administration, Washington, DC, October 2004.

Final Economic Assessment, Child Restraint Systems (FMVSS 213), Child Restraint Anchorage Systems (FMVSS 225). National Highway Traffic Safety Administration, Washington, DC, 1999.

Starnes, M. and Eigen, A. *Fatalities and injuries to 0-8 year old passenger vehicle occupants based on impact attributes.* NHTSA Technical Report No. DOT HS 809 410, National Highway Traffic Safety Administration, Washington, DC, March 2002.

Stewart D., Lang, N.J., and Emery, S. *LATCH – Lower Anchors and Tethers for Child Restraints*, Safe Ride News Publication, Seattle, 4th Edition 2005.

Weber, K. "Crash Protection for Child Passengers – A Review of Best Practice," *UMTRI Research Review,* Vol. 31, No. 3, July-September 2000, pp. 1-27.

ATTACHMENT A: GLOSSARY OF CHILD RESTRAINT TERMS

Adjuster: Hardware that can be manually changed to tighten the "fit" (attachment) of the CRS to the vehicle. Some of the types of adjusters that are commonly used are: tilt-lock, double-back, and squeeze release or push button.

Tilt-Back – The strap is released when the mechanism is tilted. (See Page E-21 of Attachment E, Training Manual for Data Collectors, for a picture of this type of adjuster.)

Double-Back or Slide – Strap is locked into place by threading it in a specific manner. The strap must be disconnected and rethreaded to lengthen or tighten it. (See Page E-21 of Attachment E, Training Manual for Data Collectors, for a picture of this type of adjuster.)

Squeeze Release or Push Button – The strap is released when the mechanism is squeezed or a button is pushed. (See Page E-28 of Attachment E, Training Manual for Data Collectors, for a picture of this type of adjuster.)

Attachment – The assembly (connector, adjuster, and often webbing) on a CRS that connects it to either the upper tether anchor or the lower attachment anchor. An attachment can be flexible or rigid.   (See also upper tether and lower attachment.)

Flexible Attachments – Attachments using webbing and a manual adjuster. (See Page E-21 of Attachment E, Training Manual for Data Collectors, for pictures of flexible attachments for upper tethers and Page E-27 for those for lower attachments.)

Rigid Attachments – Attachments built into the CRS base, fixed at 280 mm (11 in) apart, and equipped with a telescoping adjustment that attaches with the anchor. (See Page E-27 of Attachment E, Training Manual for Data Collectors, for a picture of a rigid attachment for lower attachments.)

Booster Seats

Belt-Positioning Booster (BPB) – A device that raises the child's seating height, to provide a better fit for the vehicle lap and shoulder belts. A BPB may be a low-back or a backless booster, or may have a high back that provides head restraint. Some BPBs come with LATCH systems (when they are combination seats that can be used as forward-facing seats or as BPBs). Some less commonly seen types of BPBs are those with tethered harnesses (Ride Ryte with E-Z ON Kid Y Harness) and those with an upper tether (Yound Sport by Recaro).

<u>Child Restraint Systems (CRS)</u> – An infant or child seat, vest, or similar device made for the purpose of reducing motor-vehicle-related injury and deaths in the event of a crash.

<u>Child Safety Seat (CSS)</u> – This term is generally reserved to describe infant, convertible, and forward-facing child restraint systems used with a harness. It is often used synonymously with child restraint system; however, in the pure sense, it does not encompass booster seats.

> <u>Infant-Only</u> – A rear-facing CRS for an infant. Most are for babies from 5 to 20 or 22 pounds. Some are designed with a base that stays attached to the vehicle seat, so the seat can be used as an infant carrier.

> <u>Convertible</u> – This CRS is intended for use in the rear-facing mode for a baby up to at least 20 or 22 pounds, with many of the newer ones going to 30-35 pounds. Later in the forward-facing mode, the CRS is used for toddlers over age 1 and over 22 pounds going up to 40 pounds.

> <u>Forward Facing with Harness</u> - A CRS with a full harness for toddlers over age 1, used facing forward only. Most have an upper weight limit of 40 pounds, but a few accommodate children up to 60 to 80 pounds or more.

> <u>Combination Seat</u> – A type of forward-facing child restraint that is used with an internal harness system to secure a child and then, with removal of the internal harness, is used as a high-back belt-positioning booster (BPB) seat.

<u>Connector</u> - Hardware at the end of an upper tether or lower attachments that enables the CRS to be securely fastened to an upper tether anchor or the lower attachment anchors.

> <u>Hook-On Connector</u> – A tether-type hook that is used to fasten the upper tether or lower attachments to the upper tether anchor or the lower anchors. (See p. E-27 of Attachment E, Training Manual for Data Collectors, for a picture of this type of connector.)

> <u>Push-On Connector</u> – A spring-loaded latch that automatically locks around the LATCH anchor bar when the connector is pushed directly onto the bar. This type of connector may be on a rigid or flexible attachment. (See p. E-27 of Attachment E, Training Manual for Data Collectors, for a picture of a push-on connector attached to a rigid attachment and a picture of a this type of connector attached to a flexible strap.)

<u>Integrated Seat</u> – A type of CRS which is built-into the vehicle seat, typically "camouflaged" within or under an adult seating position in the back seat. Designed for forward facing use only; a majority of them have a 5-pt harness system.

<u>LATCH</u> (Lower Anchors and Tethers for Children) is the acronym used to refer to a new system that makes child safety seat installation easier—without using seat belts. LATCH is required on most child safety seats (required by FMVSS 213) and vehicles (required by FMVSS 225)

manufactured after September 1, 2002. LATCH is not required for booster seats, car beds, and vests.

Locking Clip – A flat H-shaped metal item, used to clip seat belt webbing together, in order to secure or "lock" the vehicle seat belt around the CRS.

Lower Anchor (Vehicle) - One of a pair of horizontal bars in the area of the vehicle seat bight (space where back support and lower seat cushions meet), used in place of a vehicle seat belt to secure the base of the CRS to a vehicle with lower attachments.

Lower Attachment (Child Seat) – An assembly (connector, adjuster, and often webbing) on a child restraint that connects it to a lower anchor in the vehicle. There are three types of connector types (flexible strap with hook-on connector, flexible strap with push-on connector, and rigid attachment).

Upper or Top Tether Anchor (Vehicle) – The hardware component, such as a ring or bar, and its underlying structure in the vehicle. The upper or top tether of the child safety seat is attached to the upper or top tether anchor to secure the seat to the vehicle.

Upper or Top Tether (Child Seat) - The strap and associated hardware (connector and adjustor) that anchors the top of a CRS to the vehicle body. The upper tether strap goes from the back of the CRS to the vehicle's upper tether anchor.

"Unsure" category field in some variables. While this code does not provide useful data on upper tether or lower anchor use characteristics, it does provide information on observer's attempts to collect this data. In other words, the data was not missed, just unable to be observed and recorded.

ATTACHMENT B: SCOPE OF WORK

To reach the objectives of this project, the following task activities were performed:

1. Held an initial meeting with the Contracting Officer's Technical Representative (COTR) and other NHTSA staff to discuss project objectives and activities.

2. Finalized a work plan based on discussions from the initial meeting.

3. Developed a survey plan which identified the following: target vehicles; target population; survey type (e.g., convenience, representative); survey size; survey sites, observation form content; interview form content, and data collection procedures.

4. Conducted a meeting with the SSCs and national LATCH and CRS misuse experts to identify appropriate LATCH misuse measures for the data collection forms and identify appropriate survey procedures.

5. Developed a data collection system and coding manual that covers the vehicle form and occupant form data.

6. Conducted a pilot study to evaluate the (draft) data collection forms and data collection procedures.

7. Prepared documents (data collection forms and data collection procedures) for Office of Management and Budget (OMB) review and clearance.

8. Conducted pre-data collection activities (staff recruitment, site access and permission, training of CPS-certified staff, administrative and contractual arrangements with each SSC and their organization).

9. Conducted data collection of LATCH use and misuse, general restraint use of driver and all other occupants. Collected survey data on driver's knowledge of booster seat and LATCH issues. Collected survey data on LATCH knowledge and ease of use with drivers using LATCH systems.

10. Prepared data summary and analysis.

11. Prepared data coding manual.

12. Prepared SAS data sets.

13. Submitted draft and final reports to NHTSA.

ATTACHMENT C: OMB REVIEW

The survey plan provided supporting documentation for the Federal Government's Office of Management and Budget Review. The document contained justification information on the following: (1) reasons for collection of information; (2) purpose of data; (3) data collection techniques; (4) support for uniqueness of data (that is, has a similar data collection effort ever been done); (5) small-business impact; (6) consequences to Federal Government if study is not conducted; (7) frequency of data collection effort; (8) Federal Register notice (U.S. DOT Docket Number NHTSA –04-18816); (9) compensation to public involved in study; (10) confidentiality issues with the public; (11) sensitivity of issues in survey; (12) staff hour burden of the collection of information; (13) annual cost burden to respondents; (14) estimates of annualized cost to the Federal Government; (15) reasons for program changes; (16) publication issues from results of study; (17) display of OMB approval expiration date on forms; and (18) exceptions to the certification of paperwork reduction act.

Attachments were also included with this packet, including: the survey questionnaires and NHTSA Executive Order 12866, "Regulatory Planning and Review," which is an order requiring NHTSA to conduct periodic evaluations of the effectiveness of its Federal motor vehicle safety standards (note – the data collection supported DOT's strategic goals, i.e., safety, to promote the public health and safety by working toward the elimination of transportation-related deaths and injuries).

ATTACHMENT D: SURVEY DATA COLLECTION SITES

Data was collected in the following counties and cities in each State:

Arizona

Counties: Coconino, Maricopa, Pima, Pinal, and Yuma.
Cities: Flagstaff, Mesa, Phoenix, Sierra Vista, Tucson, and Yuma.

Florida

Counties: Duval, Hillsborough, Orange, Pinellas, and St. Johns.
Cities: Jacksonville, Orlando, St. Augustine, St. Petersburg, and Tampa.

Michigan

Counties: Clinton, Macomb, Oakland, Washtenaw, and Wayne.
Cities: Ann Arbor, Detroit, and Lansing.

Missouri

Counties: Jefferson, St. Charles, and St. Louis.
Cities: Springfield and St. Louis

North Carolina

Counties: Cabarrus, Forsyth, Guilford, and Mecklenburg.
Cities: Charlotte, Concord, Greensboro, High Point, and Winston-Salem.

Pennsylvania

Counties: Lehigh, Northampton, and Montgomery.
Cities: Allentown, Bethlehem, and West Norriton Township (Philadelphia Suburbs).

Washington

Counties: Douglas, King, Kitsap, Pierce, Snohomish, Thurston.
Cities: Everett, Federal Way, Gig Harbor, Nespalem, Olympia, Seattle, and Tacoma.

ATTACHMENT E: TRAINING MANUAL

TRAINING MANUAL

FOR

DATA COLLECTORS

Child Restraint Use Survey
(LATCH Use and Misuse)

Spring 2005

TransAnalytics, LLC.
1722 Sumneytown Pike, Box 328
Kulpsville, PA 19443

TABLE OF CONTENTS

BACKGROUND

Proper use of child restraint systems (CRSs) is the most effective way to protect young children involved in motor vehicle crashes. The National Highway Traffic Safety Administration (NHTSA) estimates that CRSs, when properly used, reduce the chance of death for infants by 71 percent, and for toddlers by 54 percent (NHTSA, 1999).

A recent NHTSA CRS misuse observation study of children weighing less than 80 pounds found that 72.6 percent of the CRSs had critical misuses which could lead to serious injury. The study found a high level of critical CRS misuse relating to vehicle seat belts. In fact, loose vehicle seat belts were observed in over 50 percent of the CRSs used by children weighing less than 40 pounds (Decina and Lococo, 2004).

NHTSA CRS misuse studies clearly show that the loose CRS attachment to the vehicle is a very serious and common misuse among the public (Decina and Knoebel, 1996; Decina and Lococo, 2004). Child passenger safety experts believe that the LATCH system can simplify CRS installation and reduce misuses of this type (Stewart and Kern, 2003).

About five years ago, NHTSA published a final rule (March 1999) establishing a uniform child restraint attachment system (CRAS) known as LATCH (Lower Anchors and Tethers for Children), (FMVSS 213, Child Restraint Systems and FMVSS 225 Child Restraint Anchorage Systems). The rule stated that all vehicles must be equipped with independent CRASs consisting of three anchorage points (two lower anchorages and one upper anchorage). Each lower anchorage consists of a bar located at the intersection of the vehicle seat cushion and seat back, in a location where passengers will not feel it. The upper anchorage is a tether anchorage. These anchorage systems are required at two rear seating positions. In addition, if a vehicle has three designated seating positions in the rear seat or a second or third row of seats, another seating position, other than the outboard position, must be equipped with a user-ready tether anchorage. This rule applies to vehicles under 8,500 pounds and buses less than or equal to 10,000 pounds.

The current technological advances in vehicle occupant protection technology, new CRS installation rulemaking and design features, and NHTSA's recognition that any manually engaged restraint system increases the possibility of incorrect CRS use, provide sufficient rationale for monitoring child restraint system use to improve passenger safety in the Nation.

Decina, L.E. and Knoebel, K.Y., 1996. Patterns of misuse of child safety seats. (Report no. DOT HS 808 440), USDOT/NHSTA, Washington, DC.

Decina, L.E. and Lococo, K.H., 2004. Misuse of child restraints. (Report no. DOT HS 809 671), USDOT/NHTSA, Washington, DC.

NHTSA, 1999. Final economic assessment, child restraint systems (FMVSS 213), child restraint anchorage systems (FMVSS 225). USDOT/NHTSA, Washington, D.C.

Stewart, D.D. and Kern, K.C., 2003. LATCH - Lower Anchors and Tethers for Child Restraints, Third Edition. Safe Ride News Publications, The Willapa Bay Company, Inc., Seattle WA.

PROJECT OBJECTIVES

The purpose of the project is to collect quantitative data and other information concerning the restraint use of children in passenger vehicles. In particular, the project will answer the following questions:

- Is LATCH being used in vehicles equipped with LATCH?

    - Are CRS tethers being used in vehicles equipped with tether anchors?
    - Are CRS lower anchor straps/connectors being used in vehicles equipped with lower anchors?

- Are children being restrained with the appropriate restraint for their age, weight, and height?

- Are children riding in the back seats of vehicles?

The study will also seek information about drivers' knowledge of booster seats and their purposes, as well as drivers' knowledge about the new LATCH technology for CRSs. In addition, if LATCH equipment is available in the vehicles, the following information will be obtained:

- Do drivers use this technology?
- What are their likes and dislikes about LATCH technology?
- Are there preferences for using seat belts to secure the CRS to the vehicle over use of lower anchors? If so, what are the reasons?

The study will also seek to identify relationships between selected driver demographic characteristics (age, gender, race/ethnicity) and restraint use of children; as well as the relationship that variables such as the demographics of other occupants in the vehicle, vehicle type, occupant seating position, and restraint type available for the seating position have on restraint use.

SURVEY DESIGN

A brief description of each survey design topic is covered. They are survey size, sampling approach, and site selection.

SURVEY SIZE

The overall target population identified for the study is *"...child passengers restrained in Child Safety Seats (CSSs) in the back seats of vehicles that are equipped with LATCH; and child passengers from birth to 12 years of age."*

The project's survey size goal is at least 1,000 child passengers from birth to age 4, restrained in CSSs in rear seating positions in passenger vehicles equipped with tethers. While collecting the data for the 1,000 quota passengers, an attempt will be made to collect child passenger restraint use data for all children from birth to age 12, as follows:

Group 1:   Child passengers from birth to age 12 (approximately 6,750); among these

Group 2:   Child passengers from birth to age 4 in CSSs in back seats (approximately 2,250); among these

Group 3:   One thousand child passengers from birth to age 4 in CSSs in back seats of vehicles equipped with tethers; and,

Group 4:   Child passengers from birth to age 4 in CSSs in back seats of vehicles equipped with lower anchors and tether anchors (LATCH) (approximately 585).

Once the data for the 1,000 child passengers are collected for Group 3, the survey can be stopped, regardless of the number of observations in the other Groups (1, 2, and 4). Group 3 target vehicles (those with tether anchors) are the highest priority during field observations.

There are seven States (Arizona, Florida, Michigan, Missouri, North Carolina, Pennsylvania, and Washington) collecting the data. Each State will collect the following data:

Group 1:   Approximately 1,000 child passengers
Group 2:   Approximately   325 child passengers
Group 3:   Approximately   150 child passengers
Group 4:   Approximately   85 child passengers

NHTSA has estimated that it will take 1.25 team hours of observation at sites with a substantial portion of late-model-year vehicles to find a Group 3 child (birth to age 4 in a CRS in the back seat of a vehicle equipped with tethers). A team will consist of two people.

NHTSA has estimated that 9 percent of all vehicles are likely to have at least one child passenger from birth to age 12; and 3 percent of vehicles will have a child passenger restrained in a CSS in the back seat. Their projections for the percentage of newer model-year (MY) vehicles that will have tether anchors and lower anchors by the data collection period of spring 2005 are as follows:

- 45 percent of vehicles will be equipped with tether anchors.
- 26 percent of vehicles will be equipped with lower anchors.

These projections for new vehicles and child safety seats (CSSs) equipped with LATCH technology are based on the following facts:

Tether anchors are installed in 80 percent of new cars manufactured on or after September 1, 1999; and 100 percent of new cars and light trucks manufactured on or after September 1, 2000. Lower anchors are installed in 20 percent of all vehicles manufactured after September 1, 2000; 50 percent of all vehicles manufactured after September 1, 2001; and 100 percent of all vehicles manufactured after September 1, 2002.

All CSSs manufactured on or after September 1, 1999, must meet new, more stringent head protection requirements. Most models of CSSs will use tethers to comply. All CSSs will have lower anchors if they have been manufactured on or after September 1, 2002.

SAMPLING APPROACH

A convenience sampling approach will be used in collecting CSS and LATCH misuse data from the targeted group of the general public. The data collection activity will receive no prior public announcement or advertisement. The approach will involve using sites such as local/community-based shopping centers, large pediatric centers, retail stores for infant and toddler products (e.g., Babies R Us), fast-food restaurants, and large child care centers. "Middle class" and "upper-middle class" communities will be used to increase the likelihood of observing newer model vehicles which would most likely be LATCH-equipped.

Target drivers and young child occupants in newer vehicles will be stopped and approached at designated safe areas at these sites and asked to participate. Upon their agreement to participate, they will be directed to a designated parking area and told to park their vehicles and turn their cars off. Past observation studies of this nature have shown less than a 10-percent refusal rate from drivers. (See *Data Collection Procedures*.)

In some States, the opportunity will exist to involve local law enforcement agencies to participate in the survey and help "recruit" target vehicles. Past efforts of this nature have used a "safety checkpoint" approach. This method involves the police slowing down every vehicle traveling through a checkpoint along a road or street. Drivers in target vehicles with young child occupants meeting the project criteria are stopped and asked to participate in the survey. Upon permission, the drivers are directed towards a designated parking area where the observations and data collection take place. This approach has been used in similar field observation studies in local communities (e.g., Upper Darby Township, Delaware County) and cities (Bethlehem, Carlisle) in Pennsylvania.

SITE SELECTION

Candidate sites will be selected across five counties and two cities in each State. The sites will include the following: in Arizona, counties in and near Phoenix and Tucson; in Florida, counties in and near Ft. Lauderdale and Miami; in Michigan, counties in and near Detroit; in Missouri, counties in and near Jefferson City and St. Louis; in North Carolina, counties in and near Charlotte and Durham; in Pennsylvania, counties in and near Bethlehem, Harrisburg, and York; and in Washington, counties in and near Seattle, Spokane, and Tacoma.

Sites used for the study will be based on several factors relating to permission and cooperation from site owners; socioeconomic characteristics of target group; high volume of target group (i.e., parents who are driving late-model vehicles with young child occupants); and physical characteristics and safety aspects of site.

Permission

Site and community cooperation are critical. Permission will need to be obtained from candidate site property managers or owners. For sites under community jurisdiction, government and police agencies will need to be notified. Your experience in conducting safety programs in many of these communities and in their shopping centers and health centers will provide some great leads to find sites. Use your network of CPS folks.

Socioeconomic Characteristics of Target Group

You should try to locate your sites in areas where the socioeconomic characteristics of the community will present a generous proportion of drivers in late-model vehicles transporting young children. After all, the main purpose of the study is to look at LATCH systems in vehicles with young child occupants. Finding these sites may require some scouting.

High Volume of Target Group

In addition, sites must have a high volume of the target group (i.e., drivers/parents in late-model vehicles with young child occupants). Shopping centers with anchor stores that are either supermarkets, large pharmacies, or household-product oriented are usually very productive. In addition, anchor stores that carry merchandise for infants and young children (e.g., Toys R Us, Babies R Us) can also be very worthwhile.

Physical Characteristics and Safety Aspects

While a large volume of the target group traveling through your site is very important in terms of efficient use of project resources, a mall-size or very large shopping center is quite unwieldy for data collection. Vehicles are traveling in all directions at various speeds. This can be very stressful and unsafe when trying to spot, slow down, and stop drivers. A site with a limited number of exits and entrances can provide a more focused ability to spot, slow down, and stop candidate drivers. Being able to easily spot all moving cars at a site will also provide a safer environment for data collection.

DEFINITIONS OF CRS AND LATCH USE AND MISUSE

Definitions of restraint system type, child harnessed/restrained, observable proper tether use, and observable proper lower anchor use were developed with the State Site Coordinators and a LATCH expert at a meeting in Philadelphia in March 2004.

Restraint System Types

Child Restraint Systems (CRS)

Infant-only with base
Infant-only without base
Convertible (used rear-facing)
Forward-facing with harness (FF only, Convertible FF, Combination seat with harness)
Integrated
Lap Top
Special Needs
Vest and Harness

Booster

Belt-positioning booster (BPB)
BPB with LATCH attachment (combination seat with harness removed)
Booster with tethered harness (e.g., *Ride Ryte with E-Z-ON Kid-Y Harness*)
Shield Booster
Integrated

Seat Belt

Lap and Shoulder
Lap only
Shoulder only

Child Harnessed/Restrained

Harness or seat belt must be buckled; and
Harness or seat belt must be over the shoulder(s); and
Harness or seat belt must be snug, with no slack, and meet the pinch test (i.e., *cannot pinch the strap to make a fold in the webbing).*

Shield booster must be secured with seat belt.

CRS or Booster attached to vehicle

CRS attached to vehicle seat with seat belt, tether, lower anchors, or any combination of the three.

A child in a BPB should be restrained by the vehicle lap and shoulder belt. A combination seat used as a BPB should not be attached to the vehicle using the LATCH lower connectors. A tether may be present and in use for some combination seats used as a BPB. A booster with a tethered Y-harness requires use of the vehicle lap belt and tether anchor to be correct.

Observable Proper Tether Use

Vehicle seating position is equipped with tether anchor.

CRS is equipped with tether. (In certain cases, CRS Manufacturer allows tether use for rear-facing installation. For a combination seat used as a BPB, a tether may be present and may be used for some models.

Tether strap is routed correctly (e.g., does not go over or around an adjustable head restraint).

CRS is attached to tether anchor designated for this position (e.g., not a cargo tie-down or lower anchor).

One tether strap is attached to a tether anchor (exception: pickup trucks with loops)

No more than one half-twist in webbing.

Tether strap is snug (meets the pinch test: cannot pinch the strap to make a fold in the webbing).

Strap is not overly tight (doesn't pull base off of vehicle seat cushion).

Tether anchor is certified for child's weight (can be deduced from reported weight).

Observable Proper Lower Anchor Use

Vehicle seating position is equipped with lower anchors.

CRS has lower attachments (flexible or rigid). For a combination seat used as a BPB, lower attachments may be present but should not be used.

Proper routing of lower attachment strap through CRS (infant base; FF/RF convertible mode)

CRS is attached to lower anchors designated for this seating position.

One connector per bar.

Both connectors attached to bars.

Connector installed right-side-up.

Flexible straps have no more than one half-twist.

Snug attachment (meets the one-inch rule: cannot move more than one inch forward or one inch side to side).

Angle achieved is appropriate for infant.

Lower anchor is certified for child's weight (can be deduced from reported weight).

DATA AND RECORDING FORMS

Observation data will be collected on child occupants, including type of restraint used and gross misuse, as well as vehicle LATCH installation use and misuse (specifically for children younger than 5 years of age in a CRS). In addition, interview data will be collected. Drivers will be asked a few demographic questions about themselves and their children, and their knowledge of booster seats and LATCH systems. If the LATCH systems are being used, drivers will be asked questions on reasons for use, ease of use, and choice over seat belts.

Demographic information and restraint type used we be obtained for all other vehicle occupants. Restraint types available for all vehicle seating positions will also be recorded.

Detailed instructions are presented in Appendix A of this manual for completing the two observation forms and two interview forms to be used in this study. The forms are shown in Appendix B.

Observation Forms

- NHTSA 1002 – "CRS and LATCH Use Observation Form."

- NHTSA 1002C – "Arrangement of Vehicle-Seating Positions and Occupant Restraint Equipment Available for Each Seating Position."

Interview Forms

- NHTSA 1002A – "CRS and LATCH Use Interview Form."

- NHTSA 1002B – "CRS and LATCH Use Interview Form, Question 6: Occupant Characteristics Chart."

DATA COLLECTION PROCEDURES

Topics covered include: staff recruitment and training, pre-data collection activities, and field procedures for data collection. All aspects of the data collection effort will be lead by the State Site Coordinators (SSCs) and their Field Site Managers (FSMs).

STAFF RECRUITMENT AND TRAINING

Data collectors responsible for making restraint system use and misuse observations are called observers. They must be CPS-certified and can be recruited through each SSC's personal network of CPS individuals or by using NHTSA's CPS-certified instructor/technician list to identify local candidates. Other electronic "CPS network" systems (e.g., CPS Listserve) can also be used to find and recruit candidate observers. Data collectors used in the last CRS misuse study would be ideal candidates as well.

Data collectors responsible for the initial contact with the driver and for conducting the interview are called greeters/interviewers. While they are not required to be CPS-certified, a comprehensive knowledge of child passenger safety issues is important for these individuals. They can be recruited through local newspaper advertisement or the SSC's personal network of resources.

Upon initiation of these staff recruitment activities, SSCs will review resumes and conduct interviews, and hire staff as needed to complete the target levels of data collection. Candidates for both positions will be screened by telephone or in person. The observers will be selected based on the following criteria:

- Have CPS certification for technician or instructor
- Have experience with CPS survey/data collection work
- Have familiarity with child development and CRSs
- Are easily able to understand project objectives and importance of their role
- Can meet a flexible schedule and be available for full duration of field observation work
- Project a sense of reliability, self-motivation, and resourcefulness
- Can provide references

The interviewers will be selected for their experience with the public. They will need to exhibit professionalism, enthusiasm, and attention to detail in accurately recording their survey information. The project will also require people who can endure the arduous nature of the data collection work, which may involve long hours, weekends, and standing for extensive periods of time under variable weather conditions.

Comprehensive training will be delivered to the data collectors. Training facilities will have a full selection of CRSs available for demonstration with a selection of vehicles that have LATCH systems. The SSCs on the project are all CPS-certified instructors and have extensive experience in conducting CRS misuse workshops.

Field observations will require careful attention to many aspects of CRSs, vehicle seat belt arrangements and types, LATCH systems, and other vehicle occupant protection components. Training will cover procedures necessary to properly identify LATCH and the vehicle restraint-belt type associated with an attached CRS.

The SSCs will use training material from the *Standardized Child Passenger Safety Training Program* (NHTSA, 2004) to cover all CRS misuse and vehicle restraint system issues. *The LATCH Manual* (Stewart and Kerrns, 2003) will also be used extensively. Additional training material will be used upon discretion of SSC and the preexisting experience and knowledge demonstrated by trainees.

Training in the classroom and in a parking lot will take two days. The following training topics and activities will be covered:

- Introduction of training objectives
- Distribution of training material (CRS and LATCH reference manuals)
- Overview of LATCH installations (tether anchors, lower anchor bars) in vehicles
- Overview of LATCH parts of CRSs (tether straps, lower connectors)
- Demonstration of LATCH misuses
- Instruction on LATCH and vehicle restraint system proper use
- Overview of field work responsibilities
- Instruction on recording data onto interview and observation forms
- Introduction on field interview and observation protocols
- Instruction on interacting with drivers, passengers, and target children
- Practice session involving LATCH misuse cases and recording data
- Practice session in parking lot (with vehicle and LATCH/CRS setups)
- Review of data collection procedures
- Onsite training at site location with "live" participants
- Debriefing and review of initial data collection effort

PRE-DATA COLLECTION ACTIVITIES

The following activities will be conducted prior to data collection:

- Selection of field sites and identification of observation spots
- Selection of field sites supervisors
- Introduction of staff to property managers or community representatives
- Development of field document packets (permission letters from property managers/owners and NHTSA Regional Office if necessary, SSC contact information, project summary sheet, etc.)
- Assembly of data collection forms, clipboard with "safety logo" on back

- Acquisition of safety vests, laminated photo ID cards, "safety logo" signs, traffic cones/caution tape, etc.

- Establishment of observation vantage points and safe pullover zones at each site

- Development of scheduling plan for field site coordinators and observation staff at each site over the course of the observation period

- Establishment of quality control, collection of data forms, and shipment methods

FIELD PROCEDURES FOR DATA COLLECTION

Data collection will be conducted throughout the week at each site depending on high-volume target vehicle time periods of the day. Past experience has found that the highest number of drivers with young child occupants can be obtained between 10 a.m. and 4 p.m. during the week, and during all daylight hours on the weekend. Field site supervisors will monitor the flow of target vehicles to determine if there are more efficient times to collect the data.

The protocol for conducting a survey and restraint use/misuse observation follows:

- Select a target vehicle entering the site and approach the driver.

- Identify oneself (interviewer) and partner (observer), briefly explain the purpose of the study (see script), and request permission to conduct the survey.

- Upon affirmation of permission, direct the driver to a parking spot within the designated survey area.

- While the interviewer is asking the driver survey questions, the observer is making observations and recording restraint system use/misuse, seating positions, and LATCH installation system data.

- Upon completion of the survey and observations, thank the driver. Provide CPS information. (Content of this information will be determined at the discretion of the SSC.)

- The interviewer and observer should "double check" the data recorded on the forms together for accuracy, before moving on to next target vehicle.

- Move back into position and wait for next vehicle.

Our data collection forms for CRS use/misuse studies will contain a script for interviewers and observers to follow. This is provided to standardize the data collection protocol.

"Hello, we are a community child-safety survey team. We are conducting a child passenger safety field observational study which will take about seven minutes. This study is sponsored by the National Highway Traffic Safety Administration. The study involves asking you a few questions and opening your vehicle doors to look at the restraint systems used by your child passengers. There is no need for your children to leave the vehicle. We will not touch your

children. You are more than welcome to watch as we observe your children in their restraints. Our observers will not change anything, and if errors are found, you will be directed to a contact to learn how to properly position your children in their restraints. This survey is voluntary and your answers will be kept confidential. Do we have your permission to conduct these activities?"

The observer will start observations from behind the driver; and work across the back seats if there is more than one target child. If the vehicle is a van, the observer will then move to the third row of seats. The front passenger side seat will be checked last. (See the Appendix for instructions on how to use observation forms.)

The interviewer will ask the driver specific questions. The purpose of the stop will be explained. This team member will request permission to conduct the observation; and if the response is positive, ask the driver to pull over to a designated area. At a minimum, each participating driver will to be asked the age, estimated weight, and height of the target children in the vehicle. (See the Appendix for instructions on how to use interview forms.)

Field supervisors will be responsible for supplying the data collectors with the coding forms; overseeing the methods used by the data collectors to make sure data were collected in the appropriate manner; collecting the forms; and verifying the completeness and accuracy of the data.

APPENDIX A

INSTRUCTIONS FOR INTERVIEW AND OBSERVATION FORMS

INSTRUCTIONS FOR COMPLETION OF
OBSERVATION FORMS NHTSA 1002 AND NHTSA 1002C

Observers have the responsibility for completing the following forms (in order listed):

(1) NHTSA 1002, "CRS and LATCH Use Observation Form" – Complete 1 form for each target child in the vehicle (e.g., less than age 13).

(2) NHTSA 1002C, "Arrangement of Vehicle Seating Positions and Occupant Restraint Equipment Available for each Seating Position" - Complete 1 form per target vehicle

INSTRUCTIONS FOR FORM 1002: CRS AND LATCH USE OBSERVATION FORM

One 1002 form will be completed for each child in the vehicle who is age 12 or younger. (For example, if there are three children in the vehicle age 12 or younger, then you will complete three 1002 Forms.)

Complete the information at the top of the form (form number, observer initials, date, State, and site identifier), and then move to the left column of the form to begin recording your observations.

LEFT COLUMN ITEMS

Child Position:

It is very important that you circle the seating position on the diagram to show where the child you are observing is seated. (The diagram may not be the same as the one that you use in your day-to-day seat-check events, so pay special attention to how you record the seating position for this study!) The diagram is a matrix of three rows and three columns designed to resemble the seating positions in a car that is parked **with the front of the vehicle facing to your left**.

The diagram shows the driver seating position at the bottom of the left column. Seating position 2 is the center seat in the first row of seats next to the driver (if there is a seat there), and seating position 3 is the front-seat passenger next to the window on the right side of the car (top of the left column in the diagram). The numbering continues in the middle column of the diagram for the second row of vehicle seats (the back seats in passenger cars), with seating position 4 behind the driver, seating position 5 in the center of the second row, and seating position 6 next to the window on the right side of the car. Some vans and SUVs have a third row of seats (shown in the third column of the diagram). These are numbered 7, 8, and 9 from the left side of the vehicle to the right side of the vehicle.

Available SB Type for this Position (whether or not used):

We want to know what kind of seat belt is available in this seating position, <u>whether or not the child is using it</u>. Circle either (1) Lap and Shoulder Belt; (2) Lap Belt Only; (3) Shoulder Belt Only; or (4) None. You would only circle "(4) None" if there is no seat belt installed in the seating position.

DO NOT circle "(4) None" for a child who is unrestrained but is sitting in a seating position where a lap and/or shoulder belt system is available. In this case, you would circle (1) Lap and Shoulder Belt.

If Child is in Position 2 or 3:

Circle "(1) Yes" if a passenger air bag switch is available. Air bag on-off switches are generally found on the dashboard. Circle "(2) No" if a passenger air bag on-off switch is not available.

If a passenger air bag on-off switch is available, circle either "(1) On" or "(2) Off" to indicate the status of the air bag.

Restraint Type Used:

This is where you will record the general kind of restraint that the child is using.

Circle "(1) CRS" if the child is riding in any of the following seat types: Infant-only with base; Infant-only without base; Convertible (used rear-facing); Forward-facing with harness (FF only, Convertible FF, Combination seat with harness); Integrated; Lap Top; Special Needs; or Vest and Harness.

Circle "(2) Booster" if the child is riding in any of the following seat types: Belt-positioning booster (BPB); BPB with LATCH attachments (Combination seat with harness removed); Booster with tethered harness (e.g., Ride Ryte with 86Y); Shield booster; or Integrated.

Circle "(3) Lap and Shoulder Belt" if a child is using a lap and shoulder belt system (even if the shoulder belt is behind the child's back or under the arm).

Circle "(4) Lap Belt Only" if the child is using only a lap belt. This would occur if the child is restrained in a vehicle position that is equipped with a lap only belt.

Circle "(5) Shoulder Belt Only" if the child is using only a shoulder belt. This would occur if the shoulder belt is an automatic one that goes across the child's body when the door is shut, and there is a separate lap belt that must be manually buckled, but is not buckled.

Circle "(6) Unrestrained" if the child is not using a CRS, a booster seat, or a seat belt. If a child is unrestrained, you will not complete any other observations on this form.

For Restraint Types 1-5, Is Child Harnessed/Restrained?

This is where harness-related or shoulder belt "gross" misuse will be recorded. The definition for "harnessed/restrained," is as follows:

- Harness or seat belt must be buckled; and
- Harness or seat belt must be over the shoulder(s); and
- Harness or seat belt must be snug, with no slack, and meet the pinch test (i.e., cannot pinch the strap to make a fold in the webbing); and
- Shield booster must be secured with seat belt.

(If a child is sitting in a CRS but the harness is not buckled, or the harness is not over the shoulders, or the harness is loose, you would circle "NO." Or, if the child is sitting in a booster, but the lap and shoulder belt system is not buckled, you would circle "NO." Or, if the child is using a lap and shoulder belt (it is buckled), but the shoulder belt is behind the back or under the arm, you would circle "NO." In other words, you would circle "NO" for situations where the child is at risk of ejection or serious injury if a crash were to occur.)

Note that for restraint types 3, 4, 5, or 6 (seat belt or unrestrained) you will not complete any other items on the form, following completion of the harnessed/restrained item.

If a child is in a CRS or a booster, continue completing the data elements in the first column, as follows.

If CRS or Booster, Type Used:

CRS Type:

There are four CRS types listed, from which to choose, if the child is using a CRS. These are the first four CRS types listed on the *Definitions* sheet developed for this project[1]:

(1) Infant-only with base
(2) Infant-only without base
(3) Convertible (used rear facing)
(4) Forward facing with harness (this includes FF only, Convertible FF, and Combination seat with harness).

If the child is restrained in a CRS, circle one of the 4 types listed (number 1, 2, 3, or 4). If the child is restrained in some other type of CRS (for example: Integrated, Lap Top, Special Needs, or Vest and Harness), circle "(9) Other" and fill in what type of CRS is being used.

---

[1] "Definitions of Restraint System Types and Observable LATCH Proper Use for NHTSA 'Child Restraint Use Survey" (DTNH22-03-C-06010).

Booster Type:

There are four types of boosters from which to choose, if a child is using a booster. These are the first four booster types listed on the *Definitions* sheet.

(5) Belt-positioning booster (BPB)
(6) BPB with LATCH attachments (this is a Combination seat with harness removed)
(7) Booster with tethered harness (e.g., Ride Ryte with 86Y)
(8) Shield booster

If the child is restrained in a booster, circle one of the four types listed (number 5, 6, 7, or 8). If the child is restrained in some other type of booster (for example, an integrated booster), circle "(9) Other" and fill in what type of booster is being used.

CRS/Booster Attachment to Vehicle:

If the CRS or booster is attached to the vehicle, there are nine combinations and types of seat belts (SBs) and LATCH attachments listed, from which to choose. Circle only one of the nine attachment combinations listed:

   (1) Lap belt only
   (2) Shoulder belt only
   (3) Lap and shoulder belt
   (4) Lower anchors only
   (5) Tether only
   (6) Lower anchors + tether
   (7) Lower anchors + SB
   (8) Lower anchors + tether + SB
   (9) Tether + SB

If the CRS or booster is not attached to the vehicle (e.g., it is just sitting on the vehicle seat), circle "(10) Not attached."

MIDDLE COLUMN ITEMS: TETHER – TOP PORTION OF COLUMN

Is a tether anchor present for this position?

For all CRSs and boosters, we want to know if a tether anchor is available for the seating position (whether or not it is being used).

Circle "(1) Yes" if a tether anchor is available for the seating position.

Circle "(2) No" if a tether anchor is <u>not</u> available for the seating position.

Circle "(3) Unsure" if you don't know if a tether anchor is available (e.g., an obstructed view or an inability to access a view).

Does CRS have a tether?

For all CRSs, we want to know if the CRS is tether equipped.

Circle "(1) Yes" if the CRS is tether equipped.

Circle "(2) No" if the CRS is not tether equipped.

Circle "(3) Unsure" if you don't know if the CRS is tether equipped (e.g., an obstructed view or an inability to access a view).

If the CRS does not have a tether, mark "NO" and move to the right column to complete observations about Lower Anchor Use.

If rear-facing CRS, can tether be used?

You will circle "(1) Yes" if the seat is facing the rear of the car (infant-only seat or a convertible in the rear-facing mode), and the manufacturer specifically indicates that a tether can be used in the rear-facing mode. At this time, there are only a few seats available in the United States that can be tethered in rear-facing operation[2]. These include the Britax Handle with Care™ (Infant), Safeline Sit 'n' Stroll™ (Convertible with stroller wheels), Britax Roundabout™ (Convertible), Britax Advantage™ (Convertible), and the new Britax convertible models with LATCH. Some of these models allow for a rear-facing tether that can be anchored toward the front of the vehicle (TFV or "Swedish Method"), or toward the rear of the vehicle (TRV or "Australian Method"). If specified in the instructions, either method is acceptable provided a suitable anchor is available. (Note - You won't have time to read the CRS manual, but hopefully you are familiar with the Britax and Safeline models, or can quickly check over any instructions for tether use on the seat label, if visible. The Safeline "Sit 'n Stroll" and the Britax "Roundabout" and "Advantage" are shown below).

---

[2] http://www.car-safety.org/latch.html

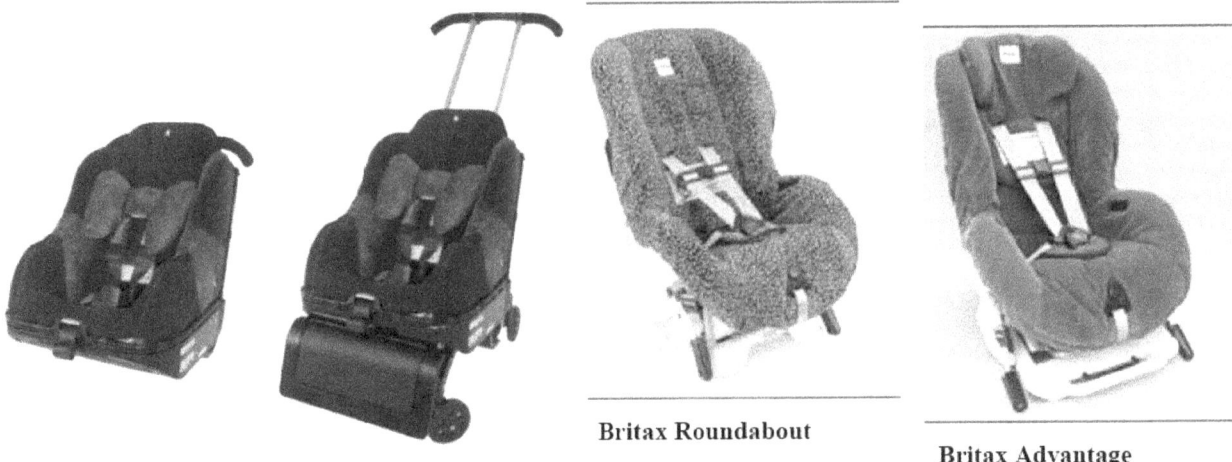

Safeline Sit 'n Stroll

Britax Roundabout

Britax Advantage

Circle (2) NO" if the seat is rear-facing, and the manufacturer does not specifically allow use of the tether in the rear-facing mode (e.g., the seat is not one of the Britax or Safeline models listed above).

Circle "(3) N/A" (stands for "not applicable") if the seat is forward-facing.

Adjustor Type:

This is the mechanism used to adjust the length of the tether strap.

(1) Tilt Lock: A strap adjustment mechanism that releases its hold on the webbing for the purpose of lengthening the strap when the hardware is held at an angle relative to the webbing; does not inhibit the strap from being shortened when the free end of the webbing is pulled. It works like a manually adjustable lap belt.

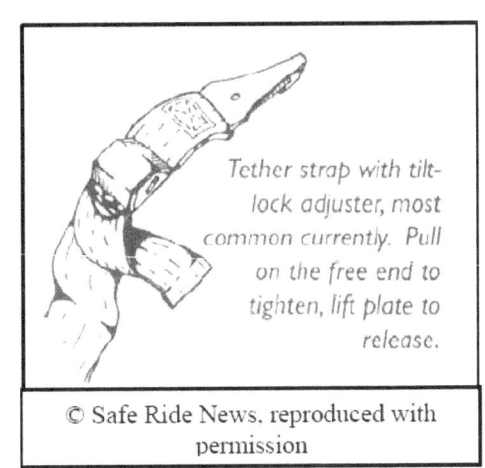

Tether strap with tilt-lock adjuster, most common currently. Pull on the free end to tighten, lift plate to release.

© Safe Ride News, reproduced with permission

(2) Double Back: Few manufacturers use this adjustment system. It uses metal slides with slots through which the strap is threaded.

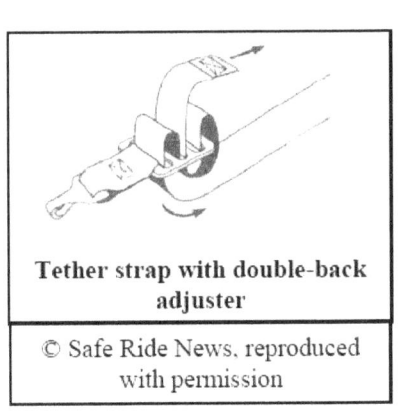

Tether strap with double-back adjuster

© Safe Ride News, reproduced with permission

(3) Other: This is any other adjustment type besides double back or tilt lock.

(4) Unsure: Circle this option if you are unsure of the adjustment type (e.g., hidden behind the seat or floor of a minivan, and you can't see it).

Is tether in use?

For CRSs that have a tether, this question relates to whether a tether is being used or not.

Circle "(1) Yes" if the tether is attached to a tether anchor or anything else in the vehicle (this question does not refer to tether misuse, which will be recorded later).

Circle "(2) No (stowed)" if the tether is not is NOT in use and it is stowed. By stowed, we are referring to the tether strap being "stored" so that it does not hang loose with the potential to cause injury during emergency maneuvering or a crash. Some CRSs have a place in the back of the seat shell to store an unused tether strap. Other seat instructions state that an unused tether strap and hook should be placed under the child restraint. If the tether strap is not in use and is stored/stowed in any manner, circle #2.

Circle "(3) No (hangs loose)" if the tether is not in use and it is hanging loose (e.g., it is a potential hazard…a child could wrap it around his neck or it could be a projectile in a crash).

NOTE: If the tether is NOT in use, move to the next column to complete the lower anchor observations. If the tether is being used, continue down the column.

MIDDLE COLUMN ITEMS: TETHER – BOTTOM PORTION OF COLUMN

Note: You will only make recordings of these items if the tether strap is in use.

Can you see what the tether is attached to?

Circle "Yes" if you can see what the tether is attached to. If you can't see what the tether is attached to with a reasonable amount of searching or minimal intrusiveness, circle "No."

Tether attachment item (1): CRS MFG permits use of RF tether.

Circle "(1) Y" if the CRS is rear facing, and it is one of the Britax or Safeline models that allows tethering of rear-facing seats.

Circle "(2) N" if the CRS is rear facing and the manufacturer is not one of the models that allows tethering in the rear-facing mode.

Circle "(3) N/A" if the CRS is forward facing.

(Note: The answer to this question should be the same as the answer to the question in the top half of the tether column "If rear-facing CRS, can tether be used?").

Tether attachment item (2): Combination used as a BPB with tether.

Circle "(1) Y" if the seat is a combination child safety seat and it is being used in the belt positioning booster (BPB) mode (harness removed, and lap and shoulder belt restraining the child).

Circle "(2) N" if the seat is a combination seat, and it is being used as a CRS (a forward-facing seat with the internal harness in use). Combination seats are used as a CRS generally for children weighing up to 40 pounds. At 40+ pounds, instructions generally state that the harness should be removed so the seat can be used as a BPB with the lap and shoulder belt.

Circle "(3) N/A" if the seat is not a combination seat.

Tether attachment item (3): Tether Routing

Items 3a-3f are designed to describe how the tether is routed. Correct routing is over a non-adjustable head restraint or under an adjustable head restraint. Use "(1) Y" for the item that describes how the tether is routed. Use "(2) N" when the item does not describe how the tether is routed, but the head restraint system described by the item is applicable to the head restraint that you are observing. Use "(3)" N/A when the item does not apply to the kind of head restraint that you are observing.

> **(3a): Tether routed over an integral/no head restraint.**
>
> Circle "(1) Y" if the tether is routed over an integral head restraint (e.g., the head restraint is built into the vehicle seat) or if there is no head restraint on the vehicle seat back.
>
> Circle "(3)" N/A if the head restraint is an adjustable head restraint, and therefore this question is not applicable to the type of head restraint observed.
>
> (NOTE: response "(2) NO" should not be used for this item)

> **(3b): Tether routed under an adjustable head restraint.**
>
> Circle "(1) Y" if the tether is routed under an adjustable head restraint.
>
> Circle "(2) N" if the head restraint is adjustable, but the tether is routed in a manner other than under the head restraint.
>
> Circle "(3) N/A" if the head restraint is integral to the seat (built in) or there is no head restraint, and therefore the question is not applicable to the head restraint being observed.

(3c): Tether routed over a raised-adjustable head restraint.

Circle "(1) Y" if the tether is routed over a head restraint that is adjustable, and is in a raised position.

Circle "(2) N" if the head restraint is adjustable, and the tether is routed in a manner other than over-the-head restraint.

Circle "(3) N/A" if the head restraint is integral to the seat (built in) or there is no head restraint, and therefore the question is not applicable to the head restraint being observed.

(3d): Tether routed over a down-adjustable head restraint.

Circle "(1) Y" if the tether is routed over a head restraint that is adjustable, and is in a lowered (full down) position.

Circle "(2) N" if the head restraint is adjustable, and the tether is routed in a manner other than over the head restraint.

Circle "(3) N/A" if the head restraint is integral to the seat (built in) or there is no head restraint, and therefore the question is not applicable to the head restraint being observed.

(3e): Tether wrapped around head restraint.

Circle "(1) Y" if the tether is wrapped around the head restraint.

Circle "(2) N" if the tether is not wrapped around the head restraint.

Circle "(3) N/A" if the head restraint is integral to the seat (built in) or there is no head restraint, and therefore the question is not applicable to the head restraint being observed.

(3f): Other routing.

If none of the items in 3a-3e describe how the tether is routed (e.g., you did not circle "(1) Y" to either 3a, 3b, 3c, 3d, or 3e), then describe here how the tether is routed.

Tether attachment item (4): Attached to designated tether anchor.

Circle "(1) Y" if the tether is attached to a tether anchor (not a cargo hook or lower anchor) and the tether anchor used is the correct one for the seating position (e.g., the anchor point located directly behind and closest to the center point of the top of the child restraint).

Circle "(2) N" if the tether is attached to a cargo hook or a lower anchor (e.g., not attached to a tether anchor) or if the tether strap is attached to a tether anchor that is designated for a different seating position. If you circle "(2) N" then use the space to describe what the tether is attached to.

Note: If you indicated earlier that you couldn't see what the tether was attached to, then leave this question blank.

Tether attachment item (5): One tether per tether anchor.

Circle "(1) Y" if only 1 tether strap (1 tether strap on 1 CRS) is attached to the tether anchor. If you are observing a pick-up truck, it is acceptable for 2 or 3 tethers (for 2 or 3 CRSs) to be attached or passed through 1 loop, so circle "(1) Y if this is the case for the seat you are observing.

Circle "(2) N" if there are 2 or more tether straps attached to the tether anchor (e.g., more that 1 CRS is using the same tether anchor, UNLESS you are observing a pickup truck, where multiple tethers may be attached to one loop.

Tether attachment item (6): Twisted no more than a half-twist.

Circle "(1) Y" if the tether strap is flat or has no more than one half-twist.

Circle "(2) N" if the tether strap has one or more twists.

Tether attachment item (7): Tether is snug (passes pinch test).

Circle "(1) Y" if the tether is snug. Snug is defined as "cannot pinch the strap to make a fold in the webbing."

Circle "(2) N" if the tether is loose (i.e., if you can pinch the strap to make a fold).

> If the tether is loose, then answer Y or N items 7a, 7b, and 7c to indicate the reason for the looseness.

Tether attachment item (8): Tether tightness allows CRS base to rest on seat cushion.

Circle "(1) Y" to indicate that the strap is not overly tight (doesn't pull base off of vehicle seat cushion).

Circle "(2) N" to indicate that the tether strap is overly tight (the base of the CRS is pulled off of the vehicle seat cushion).

Tether attachment item (9): Other

If there is anything about how the tether is attached to the vehicle that was not covered by items 1-8, circle "(1) Y" and describe the situation. If there isn't anything else about the tether attachment, circle "(2) N."

Once you complete all of the items in the middle tether column, move to the right column of the form to complete the lower anchor observations.

RIGHT COLUMN ITEMS: LOWER ANCHORS — TOP PORTION OF COLUMN

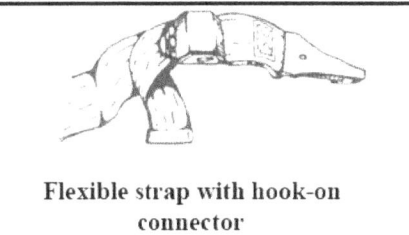

Flexible strap with hook-on connector

© Safe Ride News, reproduced with permission

Are lower anchors present in seat position?

For all CRSs and boosters, we want to know if lower anchors are available for the seating position (whether or not they are being used).

Circle "(1) Yes" if lower anchors are available for the seating position.

Circle "(2) No" if lower anchors are <u>not</u> available for the seating position.

Circle "(3) Unsure" if you don't know lower anchors are available.

Does CRS have lower attachments?

For all CRSs, we want to know if the CRS has lower attachments.

Circle "(1) Yes" if the CRS is equipped with lower attachments.

Circle "(2) No" if the CRS is <u>not</u> equipped with lower attachments.

Circle "(3) Unsure" if you don't know if the CRS is equipped with lower attachments.

If YES, Connector Type:

If you circled "(1) Yes" to the question above, then circle one of the 4 options below to indicate what type of connector is present on the lower attachments. Descriptions and figures of the connector types follow:

(1) Flexible strap, hook-on: A flexible, webbed strap with a hook (similar to a tether hook) that must be pushed in and over the bar to connect and release.

(2) Flexible strap, push-on: A flexible, webbed strap with a push-on connector (with a spring-loaded latch) that releases with a push button.

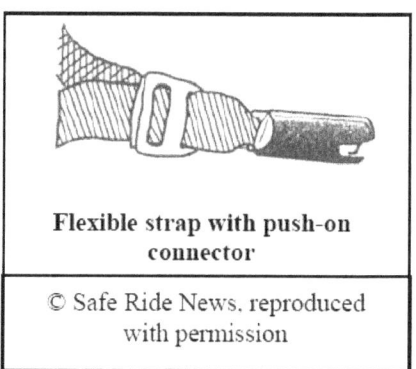

Flexible strap with push-on connector

© Safe Ride News, reproduced with permission

(3) Rigid attachment: A rigid attachment is part of the CRS frame. LATCH restraints with rigid attachments are installed by pushing the connectors onto the bars and then pushing the CRS toward the back cushion. Adjustment is automatic. Only a few CRSs currently have rigid connectors built into the base, and include Britax Expressway ISOFIX, Mercedes-Benz Duo (made by Britax and similar to the Expressway ISOFIX) and Baby Trend Latch-Loc.

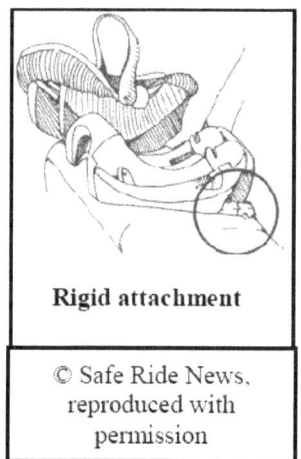

Rigid attachment

© Safe Ride News, reproduced with permission

(4) Unsure: If you are unsure of the connector type (you can't get into a position where you can see it), then circle this option.

If Flexible Strap:

Webbing Tension Release:

Flexible straps must be adjusted by the user. Circle which adjuster type is used on the flexible attachments:

(1) a squeeze release (push button)   (2) a tilt lock adjuster

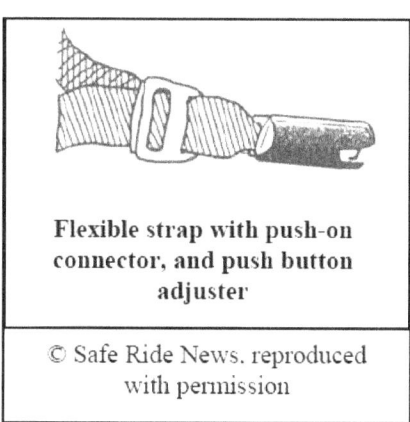

Flexible strap with push-on connector, and push button adjuster

© Safe Ride News, reproduced with permission

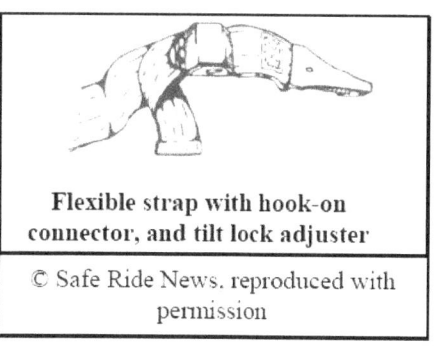

Flexible strap with hook-on connector, and tilt lock adjuster

© Safe Ride News, reproduced with permission

If you are unsure of the webbing tension release type (e.g., you can't see it), then circle "(3) Unsure."

Strap Type:

There are currently 3 types of flexible lower attachments, which are described below. Circle which type of lower attachments are present on the CRS[3]:

(1) Single – Presently, most CRSs with flexible attachments have a single strap routed through the conventional seat belt path with a hook on each end and one adjuster. For convertible CRSs, the strap must be rerouted under the pad from the rear-facing belt path to the forward-facing path. In some convertibles, the ends also must be reversed (connectors moved from one side to the other) and the connectors must be right-side up.

(2) Side Straps – A few convertibles (e.g., the discontinued Fisher-Price Safe Embrace II, Britax Marathon/Wizard) have a belt on each side that slides along the side frame of the CRS from one installation location to the other. Each strap must be tightened separately.

(3) Two Straps – Separate LATCH attachment straps for rear-facing and forward-facing positions on convertible CRSs (e.g., discontinued Cosco Triad, and an early version of the Britax Galaxy). Users must use the correct set and store the other.

---

[3] From: Stewart, D.D. and Kern, K.C. (2003). *LATCH: Lower Anchors and Tethers for Child Restraints.* Safe Ride News Publications. Seattle, WA.

If you are unsure of the strap type, circle "(4) Unsure."

Are Lower Anchors in Use?

For seating positions with lower anchors and CRSs with lower attachments, this question relates to whether the lower attachments are being used or not.

Circle "(1) Yes" if the lower attachments are attached to a lower anchor or anything else in the vehicle (this question does not refer to lower anchor misuse, which will be recorded later).

Circle "(2) No (stowed)" if the lower attachments are NOT in use and they are stowed. By stowed, we are referring to the lower attachment straps being "stored" so that they do not hang loose with the potential to cause injury during emergency maneuvering or a crash. Seat instructions variously state that unused lower attachments should be placed under the child restraint, or buckled together behind the child restraint, or clipped to storage bars on the upper sides of the seat. If the lower attachment straps are not in use and are stored/stowed in any manner, circle #2.

Circle "(3) No (hangs loose)" if the lower attachment straps are not in use and are hanging loose.

Circle "(4)" Unsure, if you can't determine whether the lower attachments are being used.

Note: If the lower attachments are not in use, or you are unsure about their use, stop the lower anchor observation for this CRS. If the lower attachments are being used, continue down the column.

RIGHT COLUMN ITEMS: LOWER ANCHORS — BOTTOM PORTION OF COLUMN

Note: You will only make recordings of these items if the lower attachments are in use.

Can you see what lower attachments are connected to?

You may have visual or tactile access to (1) what both lower attachments are connected to; (2) what only one of the lower attachments is attached to; or (3) or you may not be able to see what either attachment is connected to. Circle either (1) Both, (2) One, or (3) Neither.

Lower attachment item (1): Combination used as BPB with lower anchors

Circle "(1) Yes" if the seat is a combination seat used in the belt-positioning booster mode (the harness has been removed and the seat belt is restraining the child) and the seat is attached to the vehicle using the lower anchors.

Circle "(2) No" if the seat is a combination seat, but it is being used as a forward-facing child seat with the internal harness, and the seat is attached to the vehicle using the lower anchors.

Circle "(3) N/A" if the seat is not a combination seat.

Lower attachment item (2): Correct path used for flexible single strap:

If the lower attachment consists of a single strap:

Circle "(1) Yes" if the strap is routed in the correct belt path (e.g., toward the rear of the seat for a forward-facing seat, and to the front of the seat for a rear-facing seat).

Circle "(2) No" if the strap is <u>not</u> routed in the correct belt path.

Circle "(3) N/A" if the lower attachment type is not a single strap.

Lower attachment item (3): Attached to designated lower anchors:

Circle "(1) Yes" if the CRS is attached to lower anchors that are designated for this seating position.

Circle "(2) No" if the CRS is attached to something other than a designated anchor for the seating position. If the CRS is not attached using the designated lower anchors for the seating position, complete items 3a through 3d to describe what the lower attachments are connected to.

(3a): In non-designated position, attached using lower anchors.

Circle "(1) Yes" for situations such as a CRS in the back middle seat using outboard lower anchors (when the vehicle manufacturer prohibits outboard anchor use for the middle position.)

Circle "(2) No if the CRS is using designated lower anchors for the seating position.

(3b): In designated position, attached to seat material or spring, and not anchor.

Circle "(1) Yes" if the CRS is in a seating position with dedicated lower anchors, but the lower attachments are connected to something other than the actual lower anchors (e.g., the seat material or seat springs).

Circle "(2) No if the CRS is using designated lower anchors for the seating position.

(3c): In designated position, only one side attached to anchor.

Circle "(1) Yes" if the CRS is in a designated lower anchor seating position, but only one side of the CRS is connected to the lower anchors (e.g., only one of the two of lower connectors is in use).

Circle "(2) No if the CRS is using both designated lower anchors for the seating position.

(3d): Other:

If the CRS lower attachments are connected in any other way besides those described in 3a-3c, circle 3d and write how they are attached in the blank provided.

Lower attachment item (4): One connector per bar.

Circle "(1) Yes" if each of the 2 lower anchor bars has only one lower attachment strap connected to it.

Circle "(2) No" if one or both lower anchor bars has multiple lower attachment straps connected to it.

Lower attachment item (5): Connector installed right-side up.

Circle "(1) Yes" if both lower attachment connectors are installed to the lower anchor bar right-side up (for hooks, pushed over the bar; for push-on connectors, attached to that the release button is accessible; e.g., not wedged between the connector and the shell of the seat.)

Circle "(2) No" if one or both lower attachment connectors are installed upside down.

Circle "(3) N/A" if the CRS has rigid attachments (making upside-down connection impossible).

Lower attachment item (6): Twisted no more than a half-twist.

Circle "(1) Yes" if the lower attachments are flexible and they are attached without any twists or with only one-half of a twist.

Circle "(2) No" if the lower attachments are flexible and they are attached with a full twist or more.

Circle "(3) N/A" if the CRS has rigid attachments (cannot twist rigid attachments).

Lower attachment item (7): Tight installation (1-inch rule)

Circle "(1) Yes" if the CRS attachment is tight (e.g., it cannot move more than one inch forward or one inch side to side).

Circle "(2) No" if the CRS attachment is not tight (e.g., it moves more than one inch forward or one inch side to side).

If the CRS attachment is not tight (e.g., you circled "No" to the item above), then answer 7a-7c to indicate the cause of the loose installation:

(7a) User Error:

Circle "(1) Yes" if the loose installation is likely to be the result of human error.

Circle "(2) No" if the loose installation is likely to be the result of equipment incompatibility (e.g., the CRS base is too wide for the vehicle seat).

(7b) CRS Base Too Wide:

Circle "(1) Yes" if the loose installation is likely to be the result of the CRS base being too wide for the vehicle seat (equipment incompatibility).

Circle "(2) No" if the loose installation is likely to be the result of human error.

(7c) Other:

If the loose installation is neither likely to be the result of human error or the CRS base being too wide for the vehicle seat, then Circle 7C and write the reason in the space provided.

Lower attachment item (8): Correct angle for rear-facing CRS.

Rear-facing CRSs should be reclined at an angle between 30 and 45 degrees, with newborn and younger infants closer to 45 degrees. As infants approach 6 to 8 months, the seat may be angled closer to 30 degrees. Many rear-facing CRSs have an angle (level) gauge built into the side of the seat to verify a correct angle range.

Circle "(1) Yes" if the CRS is rear-facing, and the recline angle is between 30 and 45 degrees.

Circle "(2) No" if the CRS is rear-facing, and the angle of recline is more than 45 degrees (too reclined) or less than 30 degrees (too upright).

Circle "(3) N/A" if the CRS is a forward-facing seat.

Lower attachment item (9): Other (explain)

If you observe anything else about the lower attachment connection that is not covered by the items in 1-8 above, circle (9) and write your observations in the space provided.

INSTRUCTIONS FOR FORM 1002C:
ARRANGEMENT OF VEHICLE SEATING POSITIONS AND OCCUPANT RESTRAINT EQUIPMENT AVAILABLE FOR EACH SEATING POSITION

One 1002C form will be completed for each target vehicle, to show the seating arrangement in the vehicle, the seat belt type available for each seat, and the kind of LATCH equipment available for each seat.

TOP PORTION OF FORM

In the blocks provided at the top of the form, fill in the form number, the interviewer and observer initials, and the date. Then fill in the blank with the site description.

BOTTOM PORTION OF FORM

Enter the number of seating positions available in the vehicle, in the blank provided.

On the vehicle seating position diagram, draw an X through each seating position that is NOT available in the vehicle you are observing. The vehicle diagram is oriented so that the front of the vehicle faces the left, just like the seating position diagram shown on the 1002 form.

Circle the seat belt type for each seating position available in the vehicle, whether or not the seating position is occupied, where "L-S" indicates a lap-shoulder belt, "L" indicates a lap-only belt, and "S" indicates a shoulder-only belt.

For each tether-designated seating position, circle "T," whether or not the seating position is occupied

For each lower-anchor designated seating position. Circle "LA," whether or not the seating position is occupied.

INSTRUCTIONS FOR COMPLETION OF INTERVIEW FORMS
(NHTSA 1002A AND NHTSA 1002B)

Interviewers have the responsibility for completing the forms (in order listed):

(1) NHTSA 1002A, "CRS and LATCH Use Interview Form" – Complete front side for each target driver and on the back side, ask driver the questions for each target child (younger than 5 years old, in a CRS with a harness).

(2) NHTSA 1002B, "CRS and LATCH Use Interview Form, Question: Occupant Characteristics Chart

INSTRUCTIONS FOR FORM 1002A: CRS AND LATCH USE INTERVIEW FORM AND FOR QUESTION 6 - THE OCCUPANT CHARACTERISTICS CHART FORM (1002B)

Complete the information at the top of both forms (form number, interviewer and observer initials, date, site identifier, and time of day (AM or PM).

After initial interaction with driver (includes - introductions, explanation of survey, and permission to conduct survey with driver), proceed with survey questions on Form 1002A.

FRONT SIDE – LEFT COLUMN ITEMS

1. Vehicle Make
2. Vehicle Model
3. Vehicle Year

Ask driver this information. If driver is not certain of this information, ask permission to open driver-side door and look for sticker (on door) that identifies make, model, and year. (Note – the data entry staff will record codes.)

4. Vehicle Type

Ten categories arc provided. (Make sure data collectors know vehicle types.) Circle one.

5. Total Number of Occupants in Vehicle

Count total number of occupants. Enter number in space provided on form.

6. Occupant Characteristics *(Go to NHTSA 1002B form)*

Complete the Occupant Characteristics Chart (NHTSA 1002B) for all occupants. Ask the driver to provide the sex, age, race/ethnicity, height, and weight for each target child (age 12 or younger) <u>in each seating position</u>.

For age of the target children, drivers may respond in a variety of ways, using number of days (e.g., 14 days old), number of weeks (e.g., 2 weeks old), number of months (e.g., 24 months old), or number of years (e.g., 2 years old). Be sure to include the word or abbreviation for the unit of time provided by the driver when you record the child's age in the blank.

Weight and height for target children <u>are estimates from the driver/parent only.</u> A height chart has been provided for each team, with the four height categories color coded to help with the task of estimating height, if the parent doesn't know for sure. If the driver/parent is unable to respond, take an educated guess for weight and/or height and record the weight (in pounds) and circle the height category. (Note your initials in this box if it's your guestimate.)

Ask driver and other adults and teens, their age and race/ethnicity. If they are uncomfortable with giving actual age, provide them with age categories.

Make an observation of the restraint type being used by driver and all passengers in each seating position. Circle the restraint type being used (or unrestrained) for each occupant.

(Go back to NHTSA 1002A)

FRONT SIDE - RIGHT COLUMN ITEMS

ASK THE DRIVER THE FOLLOWING QUESTIONS IF THERE IS A CHILD LESS THAN AGE 13 IN THE VEHICLE

7. Before today, had you ever seen or heard of a "booster seat"?

Circle (1) Yes or (2) No (If "No", skip to Q. 10.)

8. For what size child should a booster seat be used?

Circle all responses *(see form)* offered by driver. (Note – Do not give them response items.) Enter numbers in the spaces provided if a driver provides a "weight, age, or height" response. If a driver gives you an answer that is not listed on the form, circle "(6) Other" and record what the driver tells you in the space provided.

9. Why is a booster seat used?

Circle all responses *(see form)* offered by driver. (Note – Do not give them response items.) If a driver gives you an answer that is not listed on the form, circle "(6) Other" and record what the driver tells you in the space provided.

10. When is it safe for your child to use a safety belt without a booster seat or child safety seat?

Circle all responses (see form) offered by driver. (Note – Do not give them response items.) If a driver gives you an answer that is not listed on the form, circle "(8) Other" and record what the driver tells you in the space provided.

11. Do you know about a new way to install a child safety seat without a safety belt?

Circle (1) Yes or (2) No (If "No", skip to Q. 13.)

12. (Upon a "yes" response) What is it called?

Circle one response (*see form*) If a driver gives you an answer that is not listed on the form, circle "(5) Other" and record what the driver tells you in the space provided.

13. Have you heard of the term "LATCH" associated with child safety seats?

Circle (1) Yes or (2) No

14. Does your <u>vehicle</u> have a place to hook the child safety seat top tether strap?

Circle (1) Yes or (2) No or (3) Don't Know

15. Does your <u>vehicle</u> have bars to attach the child safety seat bottom connectors?

Circle (1) Yes or (2) No or (3) Don't Know

On the back side of the form are questions to be asked of drivers who have target children (younger than 5 years old, in a CRS with a harness). The same questions are asked for each target child. The seating position for each target child is identified by circling the position number on the diagram for each set of questions.

BACK SIDE – TOP LEFT COLUMN

ASK THE FOLLOWING QUESTIONS FOR ALL CHILDREN IN CRSs WITH A HARNESS

16. Some child safety seats have a strap on the back of the seat near the top called a tether. Does your child safety seat have a tether?

Circle (1) Yes <u>or</u> (2) No <u>or</u> (3) Don't Know (If "No" or "Don't Know", skip to Q. 20.)

17. Are you using the tether?

Circle (1) Yes <u>or</u> (2) No (If "No", skip to Q. 19)

18. If YES, how easy or difficult is it for you to use?

Read response categories (see form) to the driver; and then circle one of five choices from (1) Very easy to (5) Very difficult. (Skip to Q. 20.)

19. If NO, why aren't you using it?

Circle as many responses as driver gives. Do not prompt answers. If a driver gives you an answer that is not listed on the form, circle "(6) Other" and record what the driver tells you in the space provided.

BACK SIDE – TOP RIGHT COLUMN

ASK THE FOLLOWING QUESTIONS FOR ALL CHILDREN IN CRSs WITH A HARNESS

20. Does your child safety seat have connectors to attach it to the vehicle?

Circle (1) Yes or (2) No or (3) Don't Know (If "No" or "Don't Know", terminate interview for this seating position.)

21. Are you using the connectors?

Circle (1) Yes or (2) No (If "No", skip to Q. 25)

22. How easy or difficult is it for you to use the connectors?

Read response categories (*see form*) to driver, and then circle one of five choices from (1) Very easy to (5) Very difficult.

23. Are you also using the safety belt?

Circle (1) Yes or (2) No (If "No", skip to Q. 26.)

24. Why do you use both the safety belt and the connectors?

Circle as many responses (*see form*) as given. Do not prompt answers. If a driver gives you an answer that is not listed on the form, circle "(3) Other" and record what the driver tells you in the space provided.

25. If NO, why aren't you using the connectors?

Driver may provide multiple answers. Do not prompt answers. Circle responses given by driver. If a driver gives you an answer that is not listed on the form, circle "(6) Other" and record what the driver tells you in the space provided.

(Interview is terminated after completion of Q. 25, since lower anchors are not being used.)

BACK SIDE – BOTTOM COLUMN

FOR LATCH (LOWER ANCHOR) USERS

26. Have you personally installed the child safety seat using the LATCH (lower anchor) system?

Circle (1) Yes <u>or</u> (2) No (If NO, terminate interview.)

27. What do you like about the LATCH (lower anchor) system?

Circle all responses (*see form*) that are given. Do not prompt answers. If a driver gives you an answer that is not listed on the form, circle "(5) Other" and record what the driver tells you in the space provided.

28. What don't you like about connecting or disconnecting the LATCH (lower anchor) system?

Circle all responses (*see form*) that are given. Do not prompt answers. If a driver gives you an answer that is not listed on the form, circle "(8) Other" and record what the driver tells you in the space provided.

29. Can you see the connection bars, or are they hidden between the seat cushions?

Circle one response (see form) for the ones given. Do not prompt answers. If a driver gives you an answer that is not listed on the form, circle "(4) Other" and record what the driver tells you in the space provided.

30. Have you had experience connecting a child safety seat to a vehicle using only the safety belts?

Circle (1) Yes <u>or</u> (2) No

31. If YES, which method do you prefer?

Circle (1) LATCH <u>or</u> (2) Safety Belts <u>or</u> (3) Undecided

31a. Why?

Write out the driver's response.

32. Is it easier to attach a child safety seat to the vehicle with the lower anchors or vehicle safety belt?

Circle (1) Lower anchors <u>or</u> (2) Vehicle safety belt <u>or</u> (3) Undecided

APPENDIX B: OBSERVATION AND INTERVIEW FORMS

US Department of Transportation
National Highway Traffic Safety Administration

CRS AND LATCH USE OBSERVATION FORM

FORM NUMBER: [　]　OBSERVER INITIALS [　]　DATE: [　] [　] [　] STATE [　] SITE [　]

Month　Day　Year

Record data for each child under age 13 on a separate form. Use blue or black ink. Circle one choice unless indicated. Use the back of this form for additional comments.

Survey Group *(Circle all that apply)*

Child age 0-12 .... Group 1

Child age 0-4, Back seat in a CRS .... Group 2

Child age 0-4, Back seat in a CRS, vehicle equipped with tether anchor ... Group 3

Child age 0-4, Back seat in a CRS, vehicle equipped with lower anchors ... Group 4

Child Position *(Circle seating position number)*:

3	6	9
2	5	8
D	4	7

← Front of vehicle points in this direction

Available SB Type for this Position (whether or not used):

(1) Lap and Shoulder Belt　　(3) Shoulder Belt Only

(2) Lap Belt Only　　(4) None

If Child is in Position 2 or 3:

Is Passenger Air Bag On-Off Switch Available?　(1) YES　(2) NO

If Available, Position of On-Off Switch:　(1) ON　(2) OFF

Restraint Type Used:

(1) CRS　　(2) Booster　　(3) Lap-Shoulder Belt

(4) Lap Belt Only　(5) Shoulder Belt Only　(6) Unrestrained

For Restraint Types 1-5, Is Child Harnessed/Restrained?

(See def. Harness or safety belt must be: buckled & over shoulders & snug.)

(1) YES　　(2) NO

---*If Restraint Type 3-6, stop observation. If Restraint type 1 or 2, continue below—*

If CRS or Booster, Type Used:

CRS Type:　　　　**Booster Type:**

(1) Infant only with Base　　(5) Belt-Positioning Booster (BPB)

(2) Infant only w/o Base　　(6) Belt-Positioning Booster w/LATCH

(3) Convertible (Rear-Facing)　(7) Booster w/tethered harness

(4) Forward-Facing w/Harness　(8) Shield Booster

(9) Other:

CRS/Booster Attachment to Vehicle:

(1) Lap belt only

(2) Shoulder belt only

(3) Lap and shoulder belt

(4) Lower anchors only

(5) Tether only

(6) Lower anchors - tether

(7) Lower anchors - SB

(8) Lower anchors - tether + SB

(9) Tether + SB

(10) Not attached

TETHER

Is a tether anchor present for this position?　(1) YES　(2) NO　(3) Unsure

Does CRS have a tether?　(1) YES　(2) NO　(3) Unsure

(If NO, stop tether observation. Move to next column for lower anchor obs)

If rear facing CRS, can tether be used?　(1) YES　(2) NO　(3) N/A

Adjustor type:　(1) Tilt Lock　(2) Double Back　(3) Other　(4) Unsure

Is tether in use?　(1) YES　(2) NO (stowed)　(3) NO (hangs loose)

(If NO, stop tether observation. Move to next column for lower anchor obs)

Can you see what the tether is attached to?　(1) YES　(2) NO

Tether attachment *(Answer all questions.)*

(1) CRS Mfg. permits use of RF tether	(1) Y	(2) N	(3) N/A
(2) Combination used as BPB with tether	(1) Y	(2) N	(3) N/A
(3a) Tether routed over an integral/no head restraint	(1) Y	(2) N	(3) N/A
(3b) Tether routed under an adjustable head restraint	(1) Y	(2) N	(3) N/A
(3c) Tether routed over a raised-adjustable head restraint	(1) Y	(2) N	(3) N/A
(3d) Tether routed over a down-adjustable head restraint	(1) Y	(2) N	(3) N/A
(3e) Tether wrapped around head restraint	(1) Y	(2) N	If NO:
(3f) Other routing.			
(4) Attached to designated tether anchor	(1) Y	(2) N	

If NO, explain:

(5) One tether per tether anchor (See def.)	(1) Y	(2) N
(6) Twisted no more than a half-twist	(1) Y	(2) N
(7) Tether is snug (passes pinch test)	(1) Y	(2) N

　　If NO:

　　(7a) User error　(1) Y　(2) N

　　(7b) Tether anchor too close to veh. seat back　(1) Y　(2) N

　　(7c) Other

(8) Tether tightness allows CRS base to rest on seat cushion	(1) Y	(2) N

(9) Other (explain):

LOWER ANCHORS

Are lower anchors present in seat position?　(1) YES　(2) NO　(3) Unsure

Does CRS have lower attachments?　(1) YES　(2) NO　(3) Unsure

If YES, Connector Type:　(1) Flexible strap, hook-on

　　(2) Flexible strap, push-on

　　(3) Rigid attachment

　　(4) Unsure

If Flexible Strap:

Webbing Tension Release: (1) Squeeze release (2) Tilt-Lock (3) Unsure

Strap Type: (1) Single　(2) Side straps　(3) Two straps　(4) Unsure

Are lower anchors in use?　(1) YES　(3) NO (hangs loose)

　　(2) NO (stowed)　(4) Unsure

(If NO or unsure, stop lower anchor observation.)

Can you see what lower attachments are connected to?

(1) Both　(2) One　(3) Neither

Lower attachment *(Answer all questions.)*

(1) Combination used as BPB with lower anchors	(1) Y	(2) N	(3) N/A
(2) Correct path used for flexible single strap	(1) Y	(2) N	(3) N/A
(3) Attached to designated lower anchors	(1) Y	(2) N	If NO:
(3a) In non-designated position, attached using lower anchors	(1) Y	(2) N	
(3b) In designated position, attached to seat material or spring	(1) Y	(2) N	

　　and not anchor

　　(3c) In designated position, only one side attached to anchor　(1) Y　(2) N

　　(3d) Other:

(4) One connector per bar	(1) Y	(2) N	
(5) Connector installed right-side up	(1) Y	(2) N	(3) N/A
(6) Twisted no more than a half-twist	(1) Y	(2) N	(3) N/A
(7) Tight installation (1-inch rule)	(1) Y	(2) N	If NO:
(7a) User Error	(1) Y	(2) N	
(7b) CRS Base Too Wide	(1) Y	(2) N	
(7c) Other			
(8) Correct angle for rear-facing CRS	(1) Y	(2) N	(3) N/A

(9) Other (explain):

NHTSA 1002

U.S. Department of Transportation
National Highway Traffic Safety Administration

CRS AND LATCH USE INTERVIEW FORM

OMB No. 2127 – 0642 (Expiration Date 04/30/08)

SITE: _____   STATE: _____

FORM NUMBER   INTERVIEWER INITIALS   OBSERVER INITIALS   TIME OF DAY [Circle one.]:   AM   PM

DATE (MO/DAY/YEAR)

"Hello. We are a community child safety survey team. We are conducting a child passenger safety field observational study which will take about 7 minutes. This study is sponsored by the National Highway Traffic Safety Administration. The study involves asking you a few questions and opening your vehicle doors to look at the restraint systems used by your child passengers. There is no need for your children to leave the vehicle. We will not touch your children. You are more than welcome to watch as we observe your children in their restraints. Our observers will not change anything, and if errors are found, you will be directed to a contact to learn how to properly position your children in their restraints. This survey is voluntary and your answers will be kept confidential. Please note that an agency may not conduct or sponsor, and a person is not required to respond to, collection of information unless it displays a currently valid OMB control number. Do we have your permission to conduct these activities?"

---

1. Vehicle Make: _____

2. Vehicle Model: _____

(Boxes for data entry use. See Vehicle Make and Model Code Manual.)

3. Vehicle Year:

4. Vehicle Type:

(1) 2-door car   (6) SUV
(2) 4-door car   (7) Pick-up truck/regular cab (2 doors)
(3) Convertible   (8) Pick-up truck/extended cab (2 doors)
(4) Mini van or van   (9) Pick-up truck/crew cab (4 doors)
(5) Station wagon   (10) Other: _____

5. Total Number of Occupants in Vehicle: _____

6. Occupant Characteristics: [Complete the Occupant Characteristics Chart for all occupants.]

ASK THE DRIVER THE FOLLOWING QUESTIONS IF THERE IS A CHILD LESS THAN AGE 13 IN THE VEHICLE

7. Before today, had you ever seen or heard of a "booster seat"?   (1) Yes   (2) No   [If No skip to Q.10.]

8. For what size child should a booster seat be used? [Circle all responses offered by driver. If "other," fill in the blank.]

(1) Weight _____   (4) For a child who has outgrown a CSS, but is too small for a SB.
(2) Age _____   (5) Don't know.
(3) Height _____   (6) Other: _____

9. Why is a booster seat used? [Circle all responses offered by driver. If "other," fill in the blank.]

(1) To make the safety belt fit the child better.   (4) Safety.
(2) It is the law.   (5) Don't know.
(3) So the child can see out of the windows.   (6) Other: _____

10. When is it safe for your child to use a safety belt without a booster seat or child safety seat?
[Circle all responses offered by driver. If "other," fill in the blank.]

(1) Weight _____   (5) When the child is tall enough that the shoulder belt doesn't cut across the chin/neck.
(2) Age _____   (6) When the child can keep his/her feet flat on the floor.
(3) Height _____   (7) Don't know.
(4) When the child's knees have   (8) Other: _____
    reached the edge of the vehicle seat.

11. Do you know about a new way to install a child safety seat without a safety belt?   (1) Yes   (2) No   [If NO, skip to Q.13].

12. What is it called?   (1) LATCH (if LATCH, also circle Yes to Q. 13 w/o asking it)   (2) Tether   (3) ISOFIX

(4) Don't know   (5) Other: _____

13. Have you heard of the term "LATCH" associated with child safety seats?   (1) Yes   (2) No

14. Does your vehicle have a place to hook the child safety seat top tether strap?   (1) Yes   (2) No   (3) Don't know.

15. Does your vehicle have bars to attach the child safety seat bottom connectors?   (1) Yes   (2) No   (3) Don't know.

NHTSA 1002A

ASK THE FOLLOWING QUESTIONS FOR ALL CHILDREN IN CRSs WITH A HARNESS

[For each child in a harnessed CRS, circle the seating position and circle the responses offered by the driver for Q 16-19 and 20-25. If response is "Other," fill in the blank.]

TETHER

3	6	9
2	5	8
D	4	7

16. Some child safety seats have a strap on the back of the seat near the top called a tether. Does your child safety seat have a tether?

(1) Yes (2) No
(3) Don't know
[If NO or Don't Know, skip to Q 20]

17. Are you using the tether?
(1) Yes (2) No
[If NO, skip to Q 19]

18. If YES, how easy or difficult is it for you to use? *[Read response categories to driver]*
(1) Very easy
(2) Relatively easy
(3) Neither easy nor difficult
(4) Somewhat difficult
(5) Very difficult

19. If NO, why aren't you using it? *[Driver may provide multiple answers]*
(1) Didn't know about it.
(2) Did not think it was important to use.
(3) Don't know how to use it
(4) Too hard to use.
(5) Rear-facing seat.
(6) Other

LOWER ANCHORS

3	6	9
2	5	8
D	4	7

20. Does your child safety seat have connectors to attach it to the vehicle?
(1) Yes (2) No (3) Don't know
[If NO or Don't Know, interview is terminated for seating position.]

21. Are you using the connectors?
(1) Yes (2) No
[If No, skip to Q 25]

22. How easy or difficult is it for you to use the connectors? *[Read response categories to driver]*
(1) Very easy
(2) Relatively easy
(3) Neither easy nor difficult
(4) Somewhat difficult
(5) Very difficult

23. Are you also using the safety belt?
(1) Yes (2) No *[If NO skip to Q 24]*

24. Why do you use both the safety belt and the connectors? *[Circle response, or fill in "Other" then skip to Q 26]*
(1) Extra secureness or safety
(2) Believed it was necessary
(3) Other

25. If NO, why aren't you using the connectors? *[Driver may provide multiple answers. Interview is terminated after completion of Q 25 as lower anchors are not used]*
(1) Didn't know about it.
(2) Did not think it was important to use.
(3) Don't know how to use it.
(4) Too hard to use.
(5) Couldn't get seat installed tightly.
(6) Other.

3	6	9
2	5	8
D	4	7

20. Does your child safety seat have connectors to attach it to the vehicle?
(1) Yes (2) No (3) Don't know
[If NO or Don't Know, interview is terminated for seating position.]

21. Are you using the connectors?
(1) Yes (2) No
[If No, skip to Q 25]

22. How easy or difficult is it for you to use the connectors? *[Read response categories to driver]*
(1) Very easy
(2) Relatively easy
(3) Neither easy nor difficult
(4) Somewhat difficult
(5) Very difficult

23. Are you also using the safety belt?
(1) Yes (2) No *[If NO skip to Q 24]*

24. Why do you use both the safety belt and the connectors? *[Circle response, or fill in "Other" then skip to Q 26]*
(1) Extra secureness or safety
(2) Believed it was necessary
(3) Other

25. If NO, why aren't you using the connectors? *[Driver may provide multiple answers. Interview is terminated after completion of Q 25 as lower anchors are not used]*
(1) Didn't know about it.
(2) Did not think it was important to use.
(3) Don't know how to use it.
(4) Too hard to use.
(5) Couldn't get seat installed tightly.
(6) Other.

FOR LATCH (LOWER ANCHOR) USERS

26. Have you personally installed the child safety seat using the LATCH (lower anchor) system? (1) Yes (2) No *[If NO, terminate interview]*

27. What do you like about the LATCH (lower anchor) system? *[Driver may provide multiple answers]*
(1) Easy to use. (2) Can see the connectors. (3) Results in a tight fit for the child safety seat. (4) N/A (I don't like LATCH) (5) Other

28. What don't you like about connecting or disconnecting the LATCH (lower anchor) system? *[Driver may provide multiple answers]*
(1) Hard to use. (2) Hard to see the bars. (3) Hard to find the bars. (4) Hard to hook the CSS to the bars. (5) Hard to release the CSS from the bars.
(6) Can't get the CSS tight. (7) N/A (I don't dislike anything about LATCH) (8) Other.

29. Can you see the connection bars, or are they hidden between the seat cushions? *[Circle one response]*
(1) Can see the connectors. (2) Hidden between the seat cushions. (3) Don't know. (4) Other:

30. Have you had experience connecting a child safety seat to a vehicle using only the safety belts? (1) Yes (2) No

31. If YES, which method do you prefer? (1) LATCH (2) Safety Belts (3) Undecided   31a. Why? (1) Lower anchors (2) Vehicle safety belt (3) Undecided

32. Is it easier to attach a child safety seat to the vehicle with the lower anchors or vehicle safety belt? (1) Lower anchors (2) Vehicle safety belt (3) Undecided

NHTSA 1062A

U.S. Department of Transportation
National Highway Traffic Safety Administration

CRS AND LATCH USE INTERVIEW FORM, QUESTION 6: OCCUPANT CHARACTERISTICS CHART

OMB No. 2127 – 0642 (Expiration Date: 04/30/08)

- For all occupants, circle: sex; race/ethnicity; and restraint type used. For 4th row or cargo area (positions 10, 11, 12), attach 2nd form and edit seating positions.
- For children under age 13, enter age in the blank. For children under 1 year, add "m" for months (e.g. 8 m). Enter child's weight and circle height category.
- For occupants age 13+, circle age category.
- CRS=child restraint system; L-S = Lap-Shoulder; L=Lap Only; S=Shoulder Only; U=Unrestrained.

FORM NUMBER   INTERVIEWER INITIALS

Seat. Pos.	Sex	Age	Race/Ethnicity	Restraint Type Used	Weight	Height (Show ref. chart)
D	1) M 2) F	___ ; 3) 13-19 4) 20-29 5) 30-39 6) 40-49 7) 50-59 8) 60-69 9) 70-79 10) 80+	1) Caucasian 2) African Amer 3) Hispanic/Latino 4) Asian 5) Native Amer 6) Other	1) CRS 2) Booster 3) L-S 4) L 5) S 6) U	___ (lb)	1) ≤27" 2) 28-40" 3) 41-57" 4) >57"
2	1) M 2) F	___ ; 3) 13-19 4) 20-29 5) 30-39 6) 40-49 7) 50-59 8) 60-69 9) 70-79 10) 80+	1) Caucasian 2) African Amer 3) Hispanic/Latino 4) Asian 5) Native Amer 6) Other	1) CRS 2) Booster 3) L-S 4) L 5) S 6) U	___ (lb)	1) ≤27" 2) 28-40" 3) 41-57" 4) >57"
3	1) M 2) F	___ ; 3) 13-19 4) 20-29 5) 30-39 6) 40-49 7) 50-59 8) 60-69 9) 70-79 10) 80+	1) Caucasian 2) African Amer 3) Hispanic/Latino 4) Asian 5) Native Amer 6) Other	1) CRS 2) Booster 3) L-S 4) L 5) S 6) U	___ (lb)	1) ≤27" 2) 28-40" 3) 41-57" 4) >57"
4	1) M 2) F	___ ; 3) 13-19 4) 20-29 5) 30-39 6) 40-49 7) 50-59 8) 60-69 9) 70-79 10) 80+	1) Caucasian 2) African Amer 3) Hispanic/Latino 4) Asian 5) Native Amer 6) Other	1) CRS 2) Booster 3) L-S 4) L 5) S 6) U	___ (lb)	1) ≤27" 2) 28-40" 3) 41-57"
5	1) M 2) F	___ ; 3) 13-19 4) 20-29 5) 30-39 6) 40-49 7) 50-59 8) 60-69 9) 70-79 10) 80+	1) Caucasian 2) African Amer 3) Hispanic/Latino 4) Asian 5) Native Amer 6) Other	1) CRS 2) Booster 3) L-S 4) L 5) S 6) U	___ (lb)	1) ≤27" 2) 28-40" 3) 41-57" 4) >57"
6	1) M 2) F	___ ; 3) 13-19 4) 20-29 5) 30-39 6) 40-49 7) 50-59 8) 60-69 9) 70-79 10) 80+	1) Caucasian 2) African Amer 3) Hispanic/Latino 4) Asian 5) Native Amer 6) Other	1) CRS 2) Booster 3) L-S 4) L 5) S 6) U	___ (lb)	1) ≤27" 2) 28-40" 3) 41-57" 4) >57"
7	1) M 2) F	___ ; 3) 13-19 4) 20-29 5) 30-39 6) 40-49 7) 50-59 8) 60-69 9) 70-79 10) 80+	1) Caucasian 2) African Amer 3) Hispanic/Latino 4) Asian 5) Native Amer 6) Other	1) CRS 2) Booster 3) L-S 4) L 5) S 6) U	___ (lb)	1) ≤27" 2) 28-40" 3) 41-57" 4) >57"
8	1) M 2) F	___ ; 3) 13-19 4) 20-29 5) 30-39 6) 40-49 7) 50-59 8) 60-69 9) 70-79 10) 80+	1) Caucasian 2) African Amer 3) Hispanic/Latino 4) Asian 5) Native Amer 6) Other	1) CRS 2) Booster 3) L-S 4) L 5) S 6) U	___ (lb)	1) ≤27" 2) 28-40" 3) 41-57" 4) >57"
9	1) M 2) F	___ ; 3) 13-19 4) 20-29 5) 30-39 6) 40-49 7) 50-59 8) 60-69 9) 70-79 10) 80+	1) Caucasian 2) African Amer 3) Hispanic/Latino 4) Asian 5) Native Amer 6) Other	1) CRS 2) Booster 3) L-S 4) L 5) S 6) U	___ (lb)	1) ≤27" 2) 28-40" 3) 41-57" 4) >57"

◄ Front of vehicle points in this direction.

NHTSAf002B

U.S. Department of Transportation
National Highway Traffic Safety Administration

Seating and Restraint Form

Arrangement of Vehicle Seating Positions and Occupant Restraint Equipment Available for each Seating Position
(For Observer Completion)

FORM NUMBER   INTERVIEWER INITIALS   OBSERVER INITIALS   DATE (MO/DAY/YEAR)

SITE: _____

****

Number of vehicle seating positions: _____

To show the seating arrangement in this vehicle, the safety belt type available for each seat, and the kind of LATCH equipment available for each seat (whether occupied or not), do the following:

- *Place an X in chart below if seating position is NOT available in vehicle.*
- *Safety Belt: Circle either L-S (for lap-shoulder), L (for lap only), or S (for shoulder only)for each seating position.*
- *Tether: Circle the T for each designated tether seating position.*
- *Lower Anchors: Circle the LA for each designated lower anchor seating position.*

3 L-S L S T LA	6 L-S L S T LA	9 L-S L S T LA
2 L-S L S T LA	5 L-S L S T LA	8 L-S L S T LA
D L-S L S T LA	4 L-S L S T LA	7 L-S L S T LA

← Front of vehicle points in this direction

NHTSA 1002C

E-53

ATTACHMENT F: DATA ENTRY AND CHECKS

Initial Data Checks

All of the data recorded on the interview and observation forms were checked in the field by the SSCs and the field observers. Inconsistencies and errors in recording information were mostly resolved with the data collectors on the day of the surveys by the field managers or SSCs. The LATCH configuration of each vehicle recorded in the survey was checked with the *LATCH –Lower Anchors and Tethers for Child Restraints* 4th Edition (Stewart, Lang, and Emery, 2005) by the SSCs and data analysis staff.

Each batch of data from a specific site was given a unique site number identifier (e.g., Site 1 - Gateway Shopping Center); and other batch-specific information related to site type (e.g., shopping center, pediatric center, fast-food restaurant, etc.), population density (e.g., urban, suburban, or rural), and socio-economic class (e.g., low/low-middle, middle/middle-upper, and upper). This other information was provided by the SSCs for each State. Socio-economic class information was a subjective call by the SSCs. This information was put on the form by project staff, not by the SSCs or field crew.

Secondary Data Checks

Prior to data entry, the data were reviewed by project staff for overall "data completeness" and were given a more specific check for inconsistencies in vehicle LATCH configurations (i.e., upper tether and/or lower anchor bar systems), target child positions and restraint use, and LATCH presence in target child seating position.

Vehicle LATCH Configurations (upper tether and/or lower anchor bar systems, Form 1002A)[23] were checked and compared with information from the most comprehensive guide on the topic (*LATCH 2005 – The Essential Guide*, by D. Stewart, N. Lang, and S. Emery – the "Green Book"). Revisions (or additions) were made to this data collection form, in red ink, upon finding missing or inconsistent data from the Green Book review.

Target Child Position and Child Restraint Use appear on Interview Form 1002B and Observer Form 1002. Form 1002C also had a record of vehicle restraint type for seating position and was used as an additional verification check. Inconsistencies in the data were checked (e.g., if the interviewer recorded that the child was in a booster seat and the observer recorded that the child was in a CRS). If a clear-cut decision could not be made about the type of restraint use (by looking at the CRS or Booster type, for example), then the data field was left blank.

LATCH Presence in Seat Occupied by Child appears on Observation Form 1002. Recordings were made for the presence of an upper tether and/or the presence of lower anchors in the seat positions occupied by the target child or children. Data on this form were checked

---

[23] The data collection forms are presented in Appendix B of the *Training Manual*, which is Attachment E of the report.

with Observation Form 1002C to verify consistency with the LATCH system attachments in the vehicle.

Data Entry Procedures

Data entry specialists received orientation to the content of the data fields for each data collection form and were trained by project staff regarding the proper procedure for entering the data. Accuracy of data entry was maximized by use of a data entry form in Microsoft Access, using combo boxes that have drop-down lists for coding variable choices in each data field. The selection of choices for the data entry specialist eliminated the need to remember values and minimized the chance of data entry error.

Data Entry Checks

Following entry of all data, further data checking activities were conducted. These are described below.

Range Checks. For each field, data was sorted in both ascending and descending orders to isolate entries outside of the range of values specified in the coding manual (e.g., a field containing the value of "4" when only 1-3 for Yes, No, and N/A were expected). This was rare, but when it occurred, the corresponding paper forms were pulled to determine what the correct code for the variable should have been, and the correct data was then entered into the database.

Skip Pattern Checks. Another data checking activity included isolating groups of variables that followed skip patterns to see if data was present when there should be none, based on the code to a prior variable. If data appeared in fields which should have been blank, the corresponding paper forms were pulled to see where the discrepancy was (e.g., an answer should have been coded as 1=Yes to the prior variable based on what was circled on the data form, instead of 2=No, and the following data was entered correctly; or the prior field was accurately coded as 2=No, and the interviewer should have skipped the following questions).

Data Consistency Checks. Some upper tether and lower anchor observations were relevant to specific restraint types (e.g., a rear-facing CRS or a belt-positioning booster). If a child was riding in a restraint type that was not relevant to the observation, the item should have been coded as 3 =N/A. Occasionally, observers incorrectly recorded such observations as a 1=Yes or a 2=No. These forms were pulled to double-check the coding of the seat, and the database entries were corrected.

Forms were pulled when there were discrepancies in Survey Group and CRS or Booster Type. For example, children riding in booster seats (Booster Type 5) who were coded as either Group 2, 3, or 4 were recoded as Group 1.

When CRS/Booster Attachment to vehicle was coded as a "3" (Lap and Shoulder Belt), and the upper tether and lower anchor observation columns indicated that an upper tether and/or the lower anchors were in use, Attachment was recoded to either "7" lower anchors + SB, or "8" lower anchors + upper tether + SB, or "9" upper tether + SB.

When 7a, 7b, and/or 7c were coded as "1=Yes" in the upper tether and lower anchor observation columns, and there were blanks for the other 1-2 variables in this sequence, the blanks were recoded to "2=No."

Review of Text Items. Some recoding of "Other" responses occurred for interview questions 8, 9, and 10; and 27, 28, and 29 when the response given for "Other" really fit into one of the existing categories. For example, for Question 28, if the interviewer wrote "seat moves too much" or "can't tighten enough" and response #6 ("Can't get CSS tight") was not coded as 1=Yes, than 28(6) was recoded from a "2" to a "1."

If numeric codes were entered in a character field, this indicated a position error in the entry of the data, either for earlier variables or later variables. Paper forms were pulled to determine where the code should have been entered.

Age Category 2 Children Who Were Missing Actual Age. Six children who were coded as Age Group 2 (age 1 to 12 years) were missing actual age. Five of these children were coded in a survey group (i.e., Survey Groups 2 through 4) for which the ages of the child passengers were defined as being between age 0 and 4. All five of these children were riding in forward-facing CRSs. Weights were provided for 4 of the 5 children, ranging from 23 pounds to 35 pounds. A decision was made to enter a "guestimate" of the actual ages of the 5 children in Survey Groups 2, 3, and 4 based loosely on their weight, because leaving them blank removes them from queries using "< age 5" as a criteria for inclusion.

ATTACHMENT G: ADDITIONAL DATA ANALYSES

CHILD OCCUPANTS YOUNGER THAN AGE 5 IN BOOSTER SEATS

There were 185 children younger than age 5 riding restrained in booster seats in back-seat seating positions.[24] Within this group of 185 children, there were 150 children being transported in booster seats in the back seats of vehicle that were equipped with upper tether anchors <u>somewhere in the vehicles,</u> but they were not included in Survey Group 3, because booster seats were not considered to be CRSs in this study as they have no internal harness, and were not designed to be installed with LATCH. Booster seats may have LATCH attachments when they are combination seats (i.e., seats that can be either used with the internal harnesses as forward-facing CRSs or as booster seats with the harnesses removed and the vehicle seat belt restraining the child).

Upper Tether Use

Upper tether straps were present on 22 of the 150 booster seats (15%) in vehicles with upper tether anchors, and the upper tether was in use for 5 of the booster seats with upper tether straps. Four of the 5 seats were combination seats being used as belt-positioning boosters (considered as a misuse) and 1 seat was a booster with a tethered harness (Ride Ryte with E-Z ON Kid Y Harness). These 5 children each weighed between 33 and 40 pounds.

Lower Attachment Use

Of the 185 children younger than 5 riding in booster seats in the back seats of vehicles, 111 were in vehicles equipped with lower anchors. Eight of these children were in restraints equipped with lower attachments, but none of the restraints had the lower attachments in use.

CHILD OCCUPANTS AGE 5 AND OLDER

A total of 376 children age 5 to 12 were observed in this study. Thirty-eight of these children (10%) were riding in the front seat. Of the 38 children in the front seat, 2 were unrestrained, 2 were riding in a lap-only belt, and 34 were restrained in a lap and shoulder belt.

Restraint use of the 338 children riding in the back seat was as follows:

- 4 percent were restrained in a CRS.
- 46 percent were riding in a booster seat.
- 43 percent were riding in a lap and shoulder belt.
- 4 percent were in a lap-only belt.
- 3 percent were unrestrained.

---

[24] This excludes the 8 children assigned to Survey Group 4 who were riding in combination seats being used as both a CRS with the internal harness and as a booster seat with the seat belt.

Upper Tether Use

Twenty-two children were riding in restraints equipped with upper tether straps; 17 of these children were riding in vehicles that were equipped with upper tether anchors. Of this subgroup of 17 children, 2 children (one child age 6 weighing 43 pounds in a booster seat with latch attachments and one child age 5 weighing 38 pounds in a forward-facing CRS with a harness) had the upper tethers in use.

Lower Attachment Use

Of the 338 children riding in the back seats, 18 were in restraints equipped with lower attachments; 11 of these children were riding in vehicles equipped with lower anchors. Of this subgroup of 11 children, 2 children (i.e., one child age 6 weighing 43 pounds in a booster seat with LATCH attachments, one child age 5 weighing 38 pounds in a shield booster) were in restraints attached to the vehicles with lower attachments.

APPROPRIATE RESTRAINT SELECTION FOR CHILD OCCUPANTS YOUNGER THAN AGE 13

One of the goals of the project was to determine if children younger than 13 years were riding in the vehicles' back seats and if they were being restrained with appropriate restraints for their age, weight, and height. Of the total sample of 1,728 children younger than age 13, 1,676 (97%) were riding in the back seats of vehicles, and 52 (3%) were riding in the front seats. While it is encouraging that most of the children in this sample were riding in the back seats, the high percentage may be more of an artifact of the target vehicle selection methodology (e.g., target vehicles were those with a child under age 5 in the back seats) than a true picture of how parents position their young children on a national basis.

While all children younger than 13 should ride in the back seat, NHTSA provides additional guidelines to help parents choose the correct restraint for each stage of their child's development:

1. Rear-facing infant seats are for babies from birth to at least 1 year old and up to 20 pounds. Children may continue to ride facing the rear of the car until they have reached the maximum weight and height specified by the CRS manufacturer (as some CRSs may be used in the rear-facing mode up to 35 pounds).

2. Forward-facing toddler seats are for when children have outgrown their infant seats to about age 4 and at least 40 pounds (or when they reach the upper weight or height limit of the child restraint).

3. Booster seats are for children who outgrow their forward-facing toddler seats, usually about age 4 and 40 pounds, until they are at least age 8, unless 4'9" or taller.

4. Lap and shoulder seat belts are for children who outgrow their booster seats, usually at age 8 or older or when they are 4' 9" or taller.

Table G-1 presents the type of restraint system used for groups of children with specific characteristics. Drivers were asked to indicate into which of four height categories their child occupants fell: ≤ 27 in, 28-40 in, 41-57 in, and > 57 in, as well as to report the weight of their child occupants. Children were not weighed or measured in the field.

Table G-1. Child Restraint Type by Characteristics of Children Important for Determining Appropriate Restraint Selection.
(Note: Appropriate restraint selection is denoted by an *.)

Child Characteristics	Restraint Type						Total
	Rear Facing	Forward Racing with Harness	Booster Seat (No Internal Harness)	Lap and Shoulder Belt	Lap Belt Only	Unrestrained	
Less than age 1 AND less than 20 pounds	204* (98%)	2 (1%)	0	0	1 (<1%)	2 (1%)	207
More than age 1 AND 20 to 40 pounds	33* (3%)	775* (78%)	160 (16%)	16 (2%)	3 (<1%)	4 (<1%)	991
41 pounds to 80 pounds	0	18* (5%)	181* (53%)	121 (36%)	11 (3%)	9 (3%)	340
81+ pounds and < 57 inches	0	0	1* (4%)	25 (92%)	0	1 (4%)	27
81+ pounds and ≥ 57 inches	0	0	0	10* (91%)	0	1 (9%)	11

Note: The total number of children represented in Table G-1 is less than 1,728 due to missing data for height (155 children), weight (52 children), and specific CRS general restraint type (5 children). Also, the 8 children in integrated (built-in) restraints are not included.

The children in lap and shoulder belts who were 81+ pounds and less than 57 inches ranged in age from 6 to 12 (2 were less than age 8), and ranged in weight from 82 to 106 pounds. Focusing only on the 265 children who were younger than 8 years old and weighed 41 to 80 pounds (as criteria for appropriate CRS/booster seat use), the following was observed: 7 percent were in forward-facing CRSs with harnesses; 65 percent were in booster seats; 23 percent were in lap and shoulder belts; 3 percent were in lap-only belts; and 2 percent were unrestrained.

www.ingramcontent.com/pod-product-compliance
Lightning Source LLC
Chambersburg PA
CBHW081500170526
45166CB00008B/2494